Ethnic Spatial Segregation in European Cities

This book provides the first in depth interpretation of how to understand the causes of ethnic residential segregation across Western European countries and the USA.

In many countries, ethnic minorities have obtained low quality housing and may be concentrated in certain parts of cities. This book asks to what extent ethnic segregation can be assigned to special preferences for housing and neighbourhoods among ethnic minorities. Is it the behaviour of the native majority, or is it a result of housing and urban policies? Ethnic segregation differs greatly across European countries and cities. Chapters discuss the extent to which these differences can be explained by welfare state systems, levels of immigration and the ethnic composition of minorities. The book also considers the impact of housing policy and the spatial structure of urban housing markets created by urban planning and policies.

This book will appeal to teachers, students and researchers working with segregation, urban sociology and geography. It will also be valuable to civil servants in central and local governments who are working with measures to combat ethnic segregation and its consequences.

Hans Skifter Andersen is adjunct professor at Danish Building Research Institute, Aalborg University, Denmark. He has for more than 20 years conducted research on segregation and deprived neighbourhoods and in the last 10 years worked with ethnic segregation and its causes, among others in connection with the NORFACE research programme on migration.

Routledge Studies in Urbanism and the City

Tourism and Everyday Life in the Contemporary City
Edited by Thomas Frisch, Christoph Sommer, Luise Stoltenberg and Natalie Stors

Citizenship and Infrastructure
Practices and Identities of Citizens and the State
Edited by Charlotte Lemanski

Pedagogies of Urban Mobilities
Kim Kullman

Balkanization and Global Politics
Remaking Cities and Architecture
Nikolina Bobic

Cities and Dialogue
The Public Life of Knowledge
Jamie O'Brien

The Walkable City
Jennie Middleton

Urban Restructuring, Power and Capitalism in the Tourist City
Contested Terrains of Marrakesh
Khalid Madhi

Ethnic Spatial Segregation in European Cities
Hans Skifter Andersen

For more information about this series, please visit www.routledge.com/Routledge-Studies-in-Urbanism-and-the-City/book-series/RSUC

Ethnic Spatial Segregation in European Cities

Hans Skifter Andersen

Routledge
Taylor & Francis Group
LONDON AND NEW YORK

First published 2019
by Routledge
2 Park Square, Milton Park, Abingdon, Oxon OX14 4RN

and by Routledge
605 Third Avenue, New York, NY 10017

First issued in paperback 2020

Routledge is an imprint of the Taylor & Francis Group, an informa business

British Library Cataloguing-in-Publication Data
A catalogue record for this book is available from the British Library

Library of Congress Cataloging-in-Publication Data
Names: Skifter Andersen, Hans, author.
Title: Ethnic spatial segregation in European cities /
Hans Skifter Andersen.
Description: First edition. | New York : Routledge, [2019] |
Series: Routledge studies in urbanism and the city | Includes
bibliographical references and index.
Identifiers: LCCN 2019003031 | ISBN 9781138358911 (hbk : alk. paper) |
ISBN 9780429280573 (ebk)
Subjects: LCSH: Minorities—Housing—Europe. | Discrimination in
housing—Europe.
Classification: LCC HD7288.72.E85 S55 2019 |
DDC 333.33/8094091732—dc23
LC record available at https://lccn.loc.gov/2019003031

ISBN 13: 978-0-367-72920-2 (pbk)
ISBN 13: 978-1-138-35891-1 (hbk)

Typeset in Times New Roman
by Apex CoVantage, LLC

Contents

Figures

Tables

Preface

This book is based on the research on housing and neighbourhood preferences among ethnic minorities and on ethnic segregation in which I have been engaged for more than 10 years. This research was a prolongation of my earlier research about the connection between segregation and the appearance of deprived neighbourhoods in cities.[1] However, the main source of the book is the results from the so-called NODES project (Nordic Welfare States and the Dynamics and Effects of Ethnic Residential Segregation), which was a major effort financed by the NORFACE migration programme in 2010–14. The project was a co-operation between researchers from four Nordic countries and produced a stream of publications.

I want to express my gratitude to the participants in this project who have contributed much to my understanding of the determinants of ethnic segregation and especially those who have contributed directly to my research. One of them, Timo Kauppinen, Social Policy Research Unit, National Institute for Health and Welfare, Finland, is a co-author of Chapter 2 in the book. Others have contributed with data and comments: Lina Hedman (Chapter 2), Lena Magnusson Turner and Susanne Søholt (Chapter 6) and Roger Andersson, Terje Wessel and Katja Vilkama (Chapter 7). Moreover, the group consisted of Mari Vaattovaara, Rikke Skovgaard Nielsen, Hanna Dhalmann, Emma Holmqvist and Ingar Brattbak.

Note

1 Published in my book *Urban Sores. On the Interaction between Segregation, Urban Decay and Deprived Neighbourhoods*. Ashgate 2003

1 Introduction and summary

The purpose of this chapter is to provide an understanding of ethnic segregation and the conditions that create it, and to give a review of the discussions from the main chapters of the book. I have chosen not to make references in this chapter, because they would amount to an unmanageable size. Instead is referred to the chapters where the main text with references is found.

In all countries, different social and ethnic groups tend to live in separate parts of the cities, even if the extent and character of this separation differs much across countries and cities. We talk about spatial segregation of groups. This book is about why ethnic minority groups in European cities often live separated from the ethnic majority and concentrated in specific neighbourhoods, especially why immigrants do. We will not analyse in much detail the extent and character of ethnic segregation in different countries. We will also not discuss the consequences of ethnic segregation, which is a separate and comprehensive research area. The subject of the book is restricted to uncovering explanations for why ethnic segregation has emerged, and is maintained, in European cities.

Research on ethnic segregation has been most extensive in North America, where massive immigration has had a longer history than in Europe. In the book we will refer to much of the American literature with the purpose of using the theoretical advances in the research on how immigrants have settled during the last 150 years. The focus will, however, mostly be on European research and evidence on ethnic segregation and its causes. A possible problem here is that available research, written in English, mostly is from Northern Europe, especially Great Britain, the Netherlands and the Nordic countries. Studies from Southern Europe have been found, but they are scarce. As ethnic segregation, and its causes, differ greatly from country to country, this is somewhat limiting to the scope of the book.

The book is based on reviews of a large number of studies of causes of ethnic segregation. Moreover, it disseminates the results from some recent comprehensive studies made in the Nordic countries, which in a more thorough way than earlier research document some of the basic explanations of segregation.

Ethnic segregation in Europe

While some studies on segregation in selected cities and countries can be found, literature comparing many countries and ethnic groups is scarce, sometimes only comparing a few countries and ethnic groups. The studies mostly concern the spatial segregation and concentration of specific ethnic groups. In many countries, however, ethnic minorities are concentrated in neighbourhoods containing several different groups, so-called *multi-ethnic neighbourhoods*. In these countries the most important kind of segregation is the separation of the whole group of ethnic minorities from the native majority.

A general conclusion from the literature is that ethnic segregation of single ethnic groups in Europe, with a few exceptions, is lower than the segregation of the black population in North America, but sometimes at the same level as other ethnic minorities in these countries. From the relatively few comparisons of the segregation of single groups in a larger number of cities, it can also be concluded that this differs greatly across countries, cities and ethnic groups. It has been found (see Chapter 8) that the most segregated groups were immigrants from south East Asia in British cities. Also, high segregation for some other groups has been found in some other cities, mostly in the central parts of Europe, but also a few in Northern and Southern countries. In all the studies large differences were shown between different groups within the same city. Moreover, the same ethnic groups were not segregated to the same extent in different cities. Extensive differences existed for them across different countries but also between different cities within the same country. These results can be influenced by limitations on which cities have been examined and by differences in the segregation data used, but they indicate that ethnic background, national conditions and individual characteristics of cities all influence ethnic segregation.

In most countries, however, the most important kind of segregation is between the native majority of a country and ethnic minorities in general, especially those who originate from Non-Western countries. Comparable data on this is scarce, except for a European survey asking for people's evaluation of the ethnic composition of their neighbourhood (see Chapter 8). These data indicate that ethnic segregation varies greatly across countries, and that these differences to some extent are connected to welfare state systems (see what follows and in Chapter 8). It is also shown, as should be expected, that there is some connection between ethnic concentration and the extent of immigration to the countries. Countries with a higher proportion of ethnic minorities also experience stronger ethnic concentrations. Especially former colonial powers like France, the Netherlands and Spain seem to be more segregated than should be expected. However, differences in ethnic segregation across countries cannot be explained by differences in the proportion of ethnic minorities in the countries. As will be discussed in the following part of the introduction, and in the other chapters of this book, ethnic segregation is a product of many different conditions that differ from country to country.

Determinants of ethnic segregation

Ethnic spatial segregation differs significantly from country to country, to some extent also from city to city within the same country. It is connected to social segregation, but not in a consistent way. Social inequality is a driving force in segregation, and the causes of social segregation are also important explanations for the spatial separation of ethnic minorities in cities, as it is a common fact that ethnic minorities, especially newly arrived immigrants, have lower incomes than the native majority in a country (see Chapter 8), and that such differences have an effect on segregation of minorities.

Social inequality and social segregation depend on the character of labour markets and welfare systems, which thus also are important determinants of ethnic segregation. In this context, integration policies and the extent to which ethnic minorities have been able to be economically integrated and obtain education, jobs and good incomes, have influenced their ability to choose better neighbourhoods and mix with the native majority.

As discussed in Chapter 7, social and ethnic segregation is not only a straightforward consequence of social inequality and ethnic diversity but is also influenced by housing policies and urban planning, which determines the way housing opportunities are located in cities and the extent to which there are small or large qualitative differences between neighbourhoods. Segregation is created by an interplay between social inequality and what can be called *spatial inequality*. In cities where all neighbourhoods are identical, there will be no social segregation, whatever social inequality there exists, because there are no advantages in staying in any neighbourhood compared to another. However, in cities where neighbourhoods are very different regarding social status, housing supply and environmental qualities, segregation tends to be high because high-income groups will be much more motivated to settle in high-quality and high-status neighbourhoods, leaving low-income groups behind in low quality and low-status neighbourhoods.

There is a mutual interaction between segregation and spatial inequality because the settlement pattern over time of respectively poor and wealthy people influences the development of neighbourhoods. In neighbourhoods where rich people settle, there will be investments in better housing and environment, and the social status will improve. In neighbourhoods which these people have left to the poor, investments will decline. Some of the neighbourhoods that previously had higher status but were abandoned by the middle class and left behind to poor people have experienced deterioration and diminished social status, which make them far less attractive. Sometimes, however, poorer neighbourhoods in central cities have gone through a reverse process where more well-to-do people have returned, resulting in what has been called gentrification.

The character of neighbourhoods and the people who are living there have thus often changed over time. It might then be argued that spatial inequality in cities is just a mirror of social inequality creating different housing demand from different groups. In this view spatial inequality does not influence segregation, but is a result of segregation. Our view is, however, that this is not quite in accordance with reality, as cities have been created over a long time and have left a historically determined housing market and spatial structure, which affects segregation, whereas the location of different social classes has changed a lot over time. Moreover, in many countries the development of cities has been highly regulated by national housing policies and urban planning that interfere with the natural processes of segregation.

Figure 1.1 depicts a model of the presumed determinants of ethnic segregation, which has been guiding the elaboration of this book.

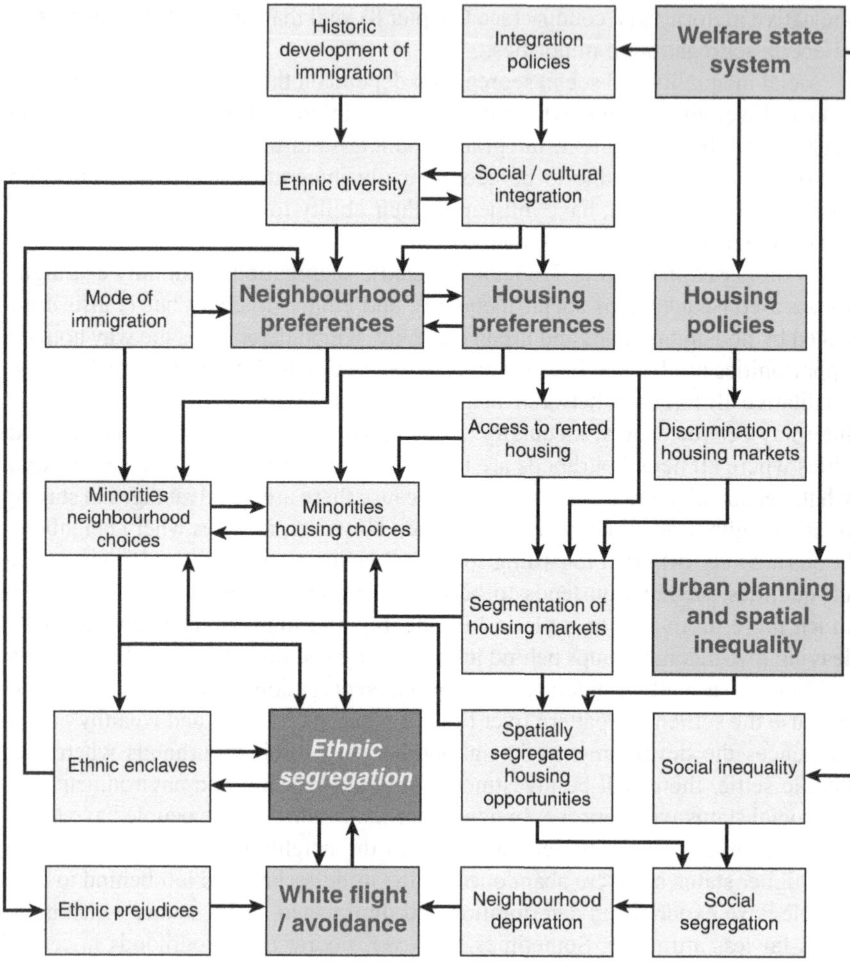

Figure 1.1 An analytic model of the determinants of ethnic segregation

The research literature suggests that ethnic segregation is not a simple consequence of lower incomes among ethnic minorities and social segregation. In short, it can be assigned to three factors: 1. Deviant preferences for housing and neighbourhoods among ethnic minorities that influence their choice of residence, 2. The strong influence of the locational preferences and moving behaviour among the native majority, who for different reasons often prefer to avoid living in neighbourhoods with ethnic minorities (*white flight* and *avoidance*), and 3. The options available for ethnic minorities on the housing markets, and the spatial distribution of different kinds of housing in cities. These three factors are, however, strongly influenced by the different societal conditions found in the actual cities and countries, which to a large extent are created by welfare state systems. Housing policies and urban planning are especially important.

In the next subsections these points will be discussed in more detail based on the review of the research literature made in the main chapters of the book.

Deviant preferences for housing among ethnic minorities (Chapter 2)

The housing preferences of ethnic minorities can to a great extent be expected to be determined by the same factors as are important for the native majority: family situation, economic resources and local housing market possibilities. It has, however, been shown in the literature that the cultural background, and the special position of immigrants in society, in some ways can influence their housing preferences and actual housing choices (Chapter 2).

One of the cultural factors is that some minorities have preferences for living in extended families, which can lead to preferences for larger homes that can be divided into parts. If this is not possible to obtain, a solution can be to find dwellings close to each other. Another cultural factor is reciprocity, the obligation to help family and friends who cannot find a dwelling. It is typical for some ethnic groups that the household often changes as members of the family move in and out. Of importance is also when and how young people leave their parents' home. This affects the housing needs of both parents and those who leave. A few studies have shown that both traditions for extended families and for how to leave parents vary between ethnic groups, and some of them also differ from the native majority.

The options of ethnic minorities in the housing market are very much determined by their preferences and potentials for purchasing their own homes. Homeownership is a much more accessible kind of housing than renting, and it is therefore important for the opportunities for choosing where to live in a city if one prefers homeownership and is able to obtain it. Economic resources are decisive, but it has been shown that some ethnic minority groups for different reasons more seldom move into homeownership than the native population no matter what incomes they have.

Buying a home demands funds for investing in property or good possibilities for obtaining a mortgage or other loans. This depends on the incomes and general economic situation of immigrants, which in general are worse than for natives. A considerable number of studies in North America and Europe have, however, shown that general household variables like income and family situation cannot alone explain why ethnic minorities in general have a lower homeownership rate than natives. It has also been found that homeownership differs between different ethnic groups and across different countries.

Cultural factors and the special position in society for immigrants have an essential importance for homeownership across countries in spite of major differences in local conditions. Some ethnic minorities have traditions that make them use their incomes and savings for other purposes than natives. Some of the labour immigrants, who came to Europe years ago, have never given up the idea of returning to their native country and have invested their savings in property there. It is, moreover, well-known that many immigrants start their own businesses in small shops that require investments. Others have family who have remained in their country of origin, and who are poor and very dependent on economic support. In this case part of the income is used for sending remittances to family members. As described in Chapter 2, another obstacle to homeownership among especially Muslim Africans are the bans for Muslims against paying interest and making profit on housing investments, which make it difficult for them to finance homeownership.

The most important reason for lower homeownership rates among ethnic minorities might be, however, that many of them are uncertain about their future and have short-term and unstable employment. It is known from housing market research that it is essential for families who decide to buy a home that they expect to have a stable or growing economy in years to come. Some studies have found that uncertainty about the future reduces the propensity of ethnic minorities to invest in housing. A study in the Nordic countries has found that the effect of increasing income on homeownership is *weaker* – or at least not stronger – among immigrants than among natives. Moreover, experiences with stable employment are important.

Immigrants' expectations about their future might depend on their social and economic integration (assimilation) in the host society. A number of studies have dealt with the connection between integration and homeownership. It is assumed that homeownership indicates long-term economic progress, plays a key role in providing financial security and it may also indicate immigrants' degree of commitment to the host country. The conclusions on this have varied. Some studies in North America have shown that length of stay in the host country only has influence on recent immigrants. A German study based on survey data has, however, indicated a connection between indicators of integration and homeownership.

Deviant preferences for neighbourhood among ethnic minorities (Chapter 3)

In all Western countries, some immigrant groups, or ethnic minorities in general, have settled together in certain neighbourhoods with a high concentration of minorities, sometimes dominated by a single ethnic group, mostly with a mix of different groups called *multi-ethnic neighbourhoods*. The existence of such neighbourhoods can to some extent be explained by special preferences and selective behaviour among ethnic minorities.

In general, preferences for location concern dimensions such as: 1. The physical environment, 2. The social and cultural environment, 3. Access to local public and private service facilities, and 4. Location and transport in a broader context. As described in Chapter 3, it has been shown that preferences vary greatly between people, and that this variation mostly depends on differences in family situation and course of life, income and other resources, and culture and life-style.

Research on locational preferences among ethnic minorities has not been very concerned with differences across ethnic groups concerning their propensity towards different kinds of physical environments or proximity to transport. Most of the research has focused on the social and cultural environment and to some extent service facilities. The absolute main focus has been on to what extent ethnic minorities prefer to live close to neighbours from their own ethnic group, subsequently if they want to avoid living in neighbourhoods dominated by the native majority or by other groups. In the literature this has often been named propensity to *self-segregation* (Chapter 3).

In principle three different kinds of preferences can be formulated: 1. To live close to family networks and friends, 2. To locate in neighbourhoods with many countrymen, or ethnic minorities with a similar culture, to become part of a local ethnic social and economic network, or 3. To avoid living close to the native majority, or other groups, to escape discrimination or racism.

The first is not very different from preferences among the native majority, but it can contribute to a concentration of ethnic minorities in certain neighbourhoods if it is more important for immigrants and if their options on the housing market are more restricted. The second is the most often mentioned motive in the literature. The third preference could be motivated by the search for security and safety in a well-known social and cultural environment and to escape what is experienced as persecution and discrimination from the native majority.

There is considerable disagreement about the importance of these preferences for segregation. As most studies have used quantitative data on the actual location of ethnic minorities, it is difficult to decide if an ethnic spatial concentration is due to preferences or to lack of options on the housing market. A few studies based on surveys and some qualitative research exist, which qualify the discussion. Results from a survey among moving ethnic minorities in Denmark indicate that the first of the three mentioned kinds of preferences is the most important.

Two different main reasons have been put forward for why some minorities prefer to settle in ethnic neighbourhoods. One is linked to having positive preferences for qualities found in these neighbourhoods that are attractive for minorities. The weaker version of this is that ethnic neighbourhoods make it possible for minorities to settle close to their tighter network: family and friends. The other has to do with a deselection of other parts of the city where minorities feel isolated, insecure or discriminated against. One can say that in some cases ethnic neighbourhoods are preferred because of their positive aspects compared to the rest of the city; in other cases they are chosen as a sanctuary for minorities feeling pursued or excluded. They could be called respectively the *ethnic enclave motive* and the *sanctuary motive*. Often the two motives are intertwined.

The sanctuary motive

Some authors in the US have argued that neighbourhood preferences of African-Americans are dominated by their concerns about the hostility they might face in white neighbourhoods. It has also been found that prejudices about other ethnic groups in the US influence neighbourhood choice. Others have proposed that preferences for avoiding other groups are intertwined with differences in social positions. If you belong to an ethnic group who usually has low incomes, you will not want to move into a neighbourhood with middle income families from another ethnic group.

The sanctuary motive has been found in some European studies in Britain, North Ireland, the Netherlands and Finland, and some of them conclude that this is the most important motive for some groups. A European survey study among ethnic minorities found a strong impact of perceived discrimination on residential location, suggesting that those exposed to or worried about discrimination were more likely to live in ethnic neighbourhoods. Some ethnic groups are more influenced by the sanctuary motive than others. Especially it has been found that immigrants from Muslim countries or from Africa more often have fear of prejudices and discrimination and want to live in ethnic neighbourhoods. Moreover, the sanctuary motive is most important for recent immigrants and minorities, who are less integrated socially and economically.

The sanctuary motive does not imply, however, that ethnic minorities in general want to live in a neighbourhood dominated by their own group or emptied of natives. Several studies show, on the contrary, that they mostly prefer a mix of natives and ethnic minorities.

The ethnic enclave motive

Positive preferences for living in ethnic neighbourhoods have been shown to depend much on the character of such urban areas. There is a mutual relationship between the character of ethnic neighbourhoods and the content and strength of preferences for living in them. Different types of ethnic neighbourhoods have been defined.

Some neighbourhoods have in the American literature been called *ghettos*, which are characterised by that the residents have not chosen to live there, but have no other choice. Ghettos are created as a consequence of strong practices of discrimination and exclusion, fostering both geographic and social isolation. Ghettos are primarily an American phenomenon and especially concern neighbourhoods with a concentration of African-Americans. It is agreed upon in the literature (see Chapter 3) that ghettos in the strict sense are not found in Europe.

Another type of ethnic neighbourhood is what is called the *ethnic enclave*. The term was originally used to characterise neighbourhoods which not only provided housing for a special ethnic group, but also employment possibilities provided by local ethnic businesses. An ideal model for this kind of ethnic enclave has been the so-called Chinatowns, inhabited by Chinese immigrants, which are mostly found in North America, but also in Britain. Earlier research claimed that such enclaves made it possible for a newcomer to live his life entirely within the confines of the community. Work, education, and access to health care, recreation and a variety of other services might be found without leaving the bounds of the ethnic economy. More recently, some have criticised these assertions, saying that employment possibilities never had such an essential influence in the ethnic enclaves.

The use of the concept ethnic enclave has later in the American literature been extended to also encompass kinds of neighbourhoods without major employment possibilities but with a high proportion of residents coming from the same ethnic group. These enclaves are culturally and to some extent economically distinct areas that are especially advantageous for certain ethnic groups, because of social, religious, cultural, commercial and linguistic services and institutions. These neighbourhoods form a framework for strong social networks for minorities with the same ethnic background and sustain cultural connections with the home country through familiar language, social networks and ethnic organisations.

Two kinds of ethnic enclaves have been defined. The first is called the *immigrant enclave model*, which is in accordance with the theories on *spatial assimilation* proposed by the Chicago School (see Chapter 4). It is assumed that this is part of new immigrants' strategy for survival and integration in their new country, and that they often have family or acquaintances in the neighbourhoods who can supply them with a social network, which can lessen their isolation, and who can support them to overcome the new challenges they meet. Housing in immigrant neighbourhoods has often a lower quality and is cheaper and more available for immigrants because housing demand from the native or white majority is low. The immigrant enclave neighbourhoods are seen as temporary starting points for an immigrant's residential career. In accordance with the spatial assimilation theory, they are expected to leave the neighbourhoods when they have been more integrated, economically and socially, in their new country.

Another type of ethnic enclave is named the *ethnic community model*, where segregated settlement can result from group preferences for establishing ethnic communities even when minorities are well integrated and able to move to more mixed neighbourhoods. These enclaves are not areas with high concentrations of poverty. They are established because the advantages of living in them might outweigh the disadvantages and are largely the product of market forces and personal choices. Earlier research in the US found that immigrant enclaves in central cities used to be the most common type of enclaves, but more recent research in North America has discovered enclaves in the suburbs that follow the ethnic community model, containing more well-to-do and integrated minorities.

The extent and character of ethnic neighbourhoods in Europe differ greatly across countries and cities. In many cases they are located in social housing estates and seen as low-status and deprived neighbourhoods. They mostly contain many different ethnic groups and do not resemble the ideal model of an ethnic enclave as described in the American literature. Neighbourhoods that most closely resemble the American ethnic communities are probably found in central cities in some countries, especially in Britain with groups from South Asia, like Pakistani and Bangladeshi. This does not mean, however, that European neighbourhoods are totally without some of the special benefits for minorities in the form of social, religious, cultural and commercial services and institutions that could make it attractive for minorities to settle there. Even if ethnic institutions are less extensive in European neighbourhoods, it could be attractive for ethnic minorities to live there because an ethnic social network is established.

The reasons for having positive preferences for ethnic neighbourhoods, stated in the somewhat scarce European studies, are similar to the reasons stated in the American literature. Most of the European studies found a desire among some members of ethnic minorities with Non-Western backgrounds for living close by others with the same ethnic background, while Western immigrants seldom had this. For many ethnic groups it is important to be part of a local community, where they can have a rich social life, feel safe and relax, and where they can get easy access to ethnic shops and facilities. It is also sometimes important that the neighbourhood provides access to amenities that facilitate religious and cultural observance, like places of worship, and to shops delivering ethnic and religious goods, like halal meat.

It has also been shown in the literature that some minorities abstain from living in ethnic neighbourhoods or directly prefer not to live there. The American literature has pointed out that an ethnic network does not necessarily need to be based on neighbourhoods, but also can be established across different parts of the city, so-called *heterolocalism*. It is argued that advances in transport and communication technology have weakened the link between spatial concentration and social network formation, resulting in the growth of ethnic communities without propinquity. Heterolocalism dismisses the relationship between spatial propinquity and ethnic communities, holding that ethnic communities are nurtured by other means than mere physical proximity. Heterolocally inclined individuals may belong to ethnic communities without having a preference for spatial overlap between residence, social relations, workplace and sites of social activities.

In some of the European qualitative studies, especially descendants of immigrants have expressed a wish to escape from ethnic enclaves because they want to escape the social control which often is present in ethnic neighbourhoods. They expressed fears of less privacy, social control, gossip or expectations of having an open door in ethnic neighbourhoods, as well as a desire for acculturation and anonymity.

Preferences for living in ethnic neighbourhoods vary greatly among ethnic minorities. The most important factor is the extent to which they are integrated/assimilated in the host country. Preferences have a strong connection with factors expressing social and economic integration, like language proficiency, social contacts with the native majority, employment, income and education. Preferences for living in neighbourhoods close to co-ethnics are thus very much determined by to what degree immigrants have been able to be an integrated part of their new society. A Danish study, however, has indicated that such factors cannot completely explain the variation in preferences for ethnic enclaves or multiethnic neighbourhoods. There are some significant differences between some of the immigrant groups, which cannot be explained by their poor measure of integration. A possible explanation could be greater feeling of discrimination and exclusion among some groups, which could lead to the sanctuary motive gaining importance. But 'cultural' explanations were also advanced. Some immigrants continue to have a strong affiliation to their country of origin and feel a split between their belonging to their new and old country, described by the concept of *diaspora*. This could influence their inclination to settle in enclaves or multi-ethnic neighbourhoods. Some ethnic groups could also have a culture that promotes a stronger wish to live in enclaves to maintain a separate cultural identity.

Spatial assimilation. The development in immigrants' locational preferences and choices over time (Chapter 4)

Ethnic segregation is not a static phenomenon. People move to and from neighbourhoods all the time, changing the ethnic and social content of them. This is to some extent connected to the changes that occur over time for individual members of ethnic minority groups. Like natives they experience changes in their housing needs and preferences during their life time that can be attributed to changes in their family situation and work career. The special condition for immigrants is, however, that at some point back in time they, or their ancestors, did arrive to a new country where many of them had poorer prerequisites for establishing a comfortable existence, mostly because of language problems and poor education, but also because of inadequate knowledge about important norms and rules and because of lack of social connections to important people and institutions. Over the course of time, sometimes generations, most immigrants have succeeded in overcoming such barriers and improved their skills for managing a better life. The most common terms used for such a development is that they have been integrated or assimilated into their new society.

The dominant theory on how immigrants settle, when they arrive, and how they change their residence over time, including following generations, is called the *spatial assimilation theory*. It has originated from the theory made by the American 'ecological' theoretical tradition of the Chicago School. The spatial location of immigrants was assumed to be closely connected with their social, cultural and economic assimilation in the host society. This theory claimed that newly arrived immigrants will tend to cluster in certain neighbourhoods, dominated by their own ethnic group, and that they over the course of time will be more dispersed in the city in pace with their integration/assimilation. This dispersal can be unleashed by two different reasons. The first is that immigrants over time will improve their economic situation, which creates a demand for better housing not found in the ethnic neighbourhoods. The other is that immigrants' preferences for living in ethnic enclaves decrease in pace with their social and cultural integration.

The spatial assimilation theory is based on a chain of assumptions:

1 That newly arrived immigrants settle in ethnic enclaves, which often are found in immigrant-dense neighbourhoods,
2 That over the course of time they will be culturally, socially and economically integrated ('assimilated' in the US literature) in their new country,
3 That this integration will change their preferences for location in ethnic enclaves, and

4 That these changed preferences will result in migration away from neighbourhoods with high concentration of ethnic minorities to neighbourhoods with better housing and environment, dominated by the native population.

All these assumptions are discussed in Chapter 4. In the research literature several of them have been doubted depending on which ethnic group or country has been studied.

Do newly arrived immigrants settle in ethnic enclaves?

The first assumption is that newly arrived immigrants prefer to settle in immigrant-dense neighbourhoods, mostly with immigrants coming from the same country or culture as themselves, sometimes called ethnic enclaves. Newly arrived immigrants have mostly been poor and have had poorer prerequisites for making a good living, because of language problems and poor education, but also because of bad knowledge about important norms and rules and lack of social connections to important people and institutions. To settle in ethnic neighbourhoods can be seen as a strategy for members of ethnic minority groups for getting support from an ethnic social network and access to ethnic services and goods; sometimes an ethnic neighbourhood might also provide better opportunities to get work through contacts to minority employers.

In the traditional literature on spatial assimilation in the US, it was found that immigrants most often settled in slum areas in the central parts of the cities. More recent research has shown, however, that some of the newer immigrants coming to North America locate in the suburbs, some in ethnic enclaves, others more dispersed, but mostly in neighbourhoods dominated by several groups of ethnic minorities. It has been suggested that one of the reasons for less settlement in the traditional ethnic enclaves might be that ethnic minorities don't any more have to live in the same neighbourhood to form a strong ethnic network, but can maintain it across different neighbourhoods, so-called heterolocalism. There has, moreover, been found major differences between the patterns of different ethnic groups.

The mode via which immigrants came to the country and the way they settled in the first place have had substantial importance for their possibilities for having success and for their choice of housing and neighbourhood. Some immigrants came as a result of what has been called chain migration, sometimes by family reunification or by having friends and family who already had become established themselves. Such immigrants have had better opportunities for success because of the support they could get from their network. Often, they also got help in finding housing, sometimes residing with family members. Other immigrants, who came as refugees or had no network to support them, have been in a much more difficult position.

The mode of immigration might have had a significant impact on the character and extent of ethnic segregation in Europe. If the earlier arrived immigrants have settled in multi-ethnic neighbourhoods, chain migration has contributed to further segregation when new immigrants have settled close to their network. For other kinds of new immigrants it might have been less obvious to settle in ethnic neighbourhoods depending on the character of these places. The initial settling of immigrants and their later residential career have depended extensively on the character of the housing market and its spatial structure. If ethnic neighbourhoods were slum areas with low quality and cheap housing, they might have been easy to access for all new immigrants. Such conditions have been found in British studies. If multi-ethnic neighbourhoods were more attractive places, like many social housing estates in Europe, access may have been difficult just after arrival, as shown in a Dutch and a Danish study.

In many European countries, governments have imposed special rules for how to settle refugees, trying to spread them across the country and across different neighbourhoods. The general policy in countries for how to receive refugees has had significant importance for how and where they settle. In some countries refugees are allocated to different municipalities according to fixed rules of geographical distribution. Sometimes local authorities use these powers to prevent refugees from settling in ethnic neighbourhoods and ensure they are spread throughout the city and countryside.

It can thus be concluded that the first assumption in the spatial assimilation theory, that immigrants at their arrival settle in ethnic enclaves, only partially can be affirmed in Europe and that it differs greatly from country to country depending on the conditions on the housing market.

Do immigrants always get integrated/assimilated in the course of time?

The second assumption in the spatial assimilation theory is that immigrants over the course of time will be integrated/assimilated in their new host country. This concerns their social and cultural integration, which might influence their propensity to prefer to have neighbours from their own ethnic group and to live in ethnic enclaves, and their integration in the economic system, which determines their financial situation and their chances to settle in more prosperous neighbourhoods together with the native majority.

The classical assimilation theory assumed that immigrants and their descendants during the course of time were integrated/assimilated culturally, socially and economically into the host society. This has been questioned in American studies of newer immigrant groups and there has been an extensive debate regarding to what extent the theory is still relevant. It has been claimed that American culture varies greatly by locality and social class and that the classical theory made a mistake in presupposing integration into the values of the white middle class. Some have formulated a theory of *segmented assimilation*, where immigrants and subsequent generations are integrated into different segments of American culture and some into permanent poverty and assimilation into the underclass, what has been coined *downward assimilation*. Results from more recent studies show still signs of assimilation during the course of time, but there are big differences between different groups. It has been shown that for some groups assimilation takes longer than for the earlier European immigrants.

In Europe the concept of integration of immigrants has had a somewhat different meaning in different countries. Immigration has in most countries had a much shorter history than in the US. The first generation of immigrants is still heavily dominating except for countries that are former colonialist powers. Europe consists of many different countries with large differences in integration contexts, which include differences in institutional arrangements in education, the labour market, housing, religion and legislation. These differences have had a significant impact on the kind and speed of immigrants' integration. It is to be expected that there is a connection between general welfare policies and efforts towards integration of immigrants. Countries with highly developed welfare policies might also use more resources on helping immigrants to adapt to their new country. It has also been discussed if the national, often negative, discourses in countries on the qualifications of immigrants and the effects of immigration in general influence integration. Some authors are of the opinion that negative attitudes by natives towards immigrants have affected integration.

Some European studies support the classical assimilation theory while others do not. Some could not support the existence of a uniform convergence process for all immigrant groups; others found that different ethnic groups established deviant subcultures, often in specific urban locations. Some studies have been able to document so-called *straight-line assimilation*: improvement in the economic integration of descendants compared to first generation immigrants.

It can be concluded that assimilation theory has some validity for most immigrants in Europe, but that integration, and the speed of it, differs much across countries and ethnic groups. There is evidence that some immigrants do not enter into mainstream society, but integrate into an ethnic underclass, what has been called downward assimilation. A similar situation has been found in the US, especially regarding immigrants with darker skin who have difficulties in getting integrated, which could be caused by more discrimination against such groups.

Do immigrants change their preferences for location over time?

The third assumption upon which the spatial assimilation theory is based is that immigrants change their preferences for housing and neighbourhood in pace with their integration over time. While there are several studies on the connection between the actual location of ethnic minorities and their integration over time, studies of changes in immigrants' stated preferences over time, or between generations, are rare. Qualitative studies among British ethnic minorities have uncovered that younger people especially, gradually want to move away from ethnic neighbourhoods in the central city into more ethnically mixed neighbourhoods in the suburbs, mainly motivated by moving for better housing, schools and neighbourhoods, but also for getting more privacy and independence from their own ethnic group. Two studies based on survey data have been found. A Dutch study showed that preferences for living with co-ethnics were lower among immigrants, who had been in the Netherlands for more than 10 years, compared to the newly arrived. A Danish study showed that immigrants' preferences for living in multi-ethnic neighbourhoods or ethnic enclaves depend on their social and economic integration. As in the US, spatial assimilation in Europe often is connected with moving into homeownership. Many studies have thus pointed to a link between homeownership and social and cultural integration/assimilation.

Does spatial assimilation actually take place?

Fourthly, it is assumed that immigrants change their residential location over time in pace with their integration and move to more ethnically mixed neighbourhoods. While earlier American studies on the settlement pattern of European immigrants confirmed the spatial assimilation theory, later studies on newer immigrants from different parts of the world, especially from Latin America, have questioned parts or all of the theory. Especially the earlier assumption that new immigrants always settle in ethnic enclaves in central city slums and later move to better housing in the suburbs has been questioned, even if some groups follow the traditional central city to suburb pattern. Some immigrants do not settle in ethnic enclaves and others move directly to ethnic enclaves in the suburbs.

Other American research has pointed to major differences across immigrants belonging to different ethnic groups concerning spatial assimilation. Some groups are very slow concerning both social and spatial assimilation, which sometimes takes several generations to go through, called *delayed assimilation.* It has been shown that racial differences have had importance. Especially immigrants with darker skin were slower to assimilate spatially, because of stronger discrimination. Some have found the concept of segmented assimilation applicable to the spatial settlement of such immigrants over time, meaning that they never leave poor and deprived multi-ethnic neighbourhoods.

Different kinds of studies have provided evidence that spatial assimilation to some extent takes place in Europe, but also that it differs greatly across ethnic groups and to some extent across countries. Several studies have shown that ethnic minorities living in multi-ethnic neighbourhoods have lower incomes than those living in other neighbourhoods. A larger European survey-based study revealed that residence in ethnic neighbourhoods tended to decrease with the passage of time in the host country, and from the data concluded that immigrants tend to translate their improved economic resources into changed residential location. This pattern was found to be quite uniform across the countries, despite substantial differences in social, political and economic structure, welfare policy, housing policy and immigrant integration policy.

Studies comparing ethnic minorities moving to and from multi-ethnic neighbourhoods have shown that income and education have a strong effect. While in-movers most often are unemployed and have low education and income, those who move out have spent a longer time in the country and have better income and employment. Other studies, based on data on how ethnic minorities relocate between different kinds of neighbourhoods, have, however, shown differing results. Some have found positive effects of income and education for change of neighbourhood, implying spatial assimilation. Others have only found a weak connection between income and neighbourhood improvements among moving families, which could insinuate that many ethnic minority families make their housing career within certain neighbourhoods. Some studies have found that many residents in ethnic neighbourhoods have remained there for a long time and that most movers had re-housed within their neighbourhood, while others have found that ethnic minorities do not stay longer in multi-ethnic neighbourhoods than in other neighbourhoods.

The few studies based on longitudinal data, following immigrants' residential career from their arrival and in following years, indicate that they in the first years of their stay increasingly move into multi-ethnic neighbourhoods, which is the opposite of spatial assimilation. This can properly be explained by their very restricted options in the housing market just after arrival, which forces them into an unsatisfying housing situation. The easiest way for them to improve their housing in the first years is to move into multi-ethnic neighbourhoods where housing is easier to obtain, either because of low quality or because demand from the native majority is low. The only study following immigrants' housing career more than five years, a Danish study, showed that the presence in multi-ethnic neighbourhoods declined after 10 to 15 years of stay in the country, pointing to some degree of spatial assimilation. This development was found also when controlled for income changes, which indicate that social integration and changed preferences have a separate effect.

A couple of studies from the Netherlands and Denmark have looked at differences in the settlement pattern between first and second-generation immigrants, so-called *straight line assimilation*. They found that the second generation moved less frequently into multi-ethnic neighbourhoods than the first generation and that young people leaving home settled more dispersed than their parents.

It has been shown that ethnic segregation and presence in multi-ethnic neighbourhoods differ greatly across ethnic groups. It is not well researched, however, to what extent this is due to cultural differences or to duration of stay and patterns of social and economic integration. A European study showed that different ethnic and cultural subgroups experienced divergent patterns of residential attainment, even after taking into consideration the differences among groups in length of residence in the host country and in socio-economic characteristics. Differences in spatial assimilation behaviour across ethnic groups have also been found in some of the other studies. The Danish study described indicated that immigrants coming from different parts of the world to some extent followed the same pattern of spatial assimilation, but at a different level.

The importance of locational preferences and moving behaviour among the native majority (Chapter 5)

Residential segregation is created by the behaviour of individual households, who move across neighbourhoods inside and across cities. This behaviour is a result of individual residential preferences and options on the housing market. Ethnic segregation is not only a result of the behaviour of ethnic minorities but also of the behaviour of the native-born population. The question is: which kind of moving behaviour has the dominant importance for the creation and maintenance of ethnic segregation and multi-ethnic neighbourhoods? Is it the behaviour of ethnic minorities or the behaviour of natives? Is it because ethnic minorities more often move into multi-ethnic neighbourhoods or because they tend to remain there when they have gotten access? Or is it because natives move away when ethnic minorities move in, or because they do without choosing these neighbourhoods?

The creation and maintenance of multi-ethnic neighbourhoods has been the result of what has been called ethnic selective migration, ethnic differences between those leaving, entering and staying in the neighbourhoods. Four different kinds of selective moves can be defined. When the native population more often move away from neighbourhoods with an ethnically mixed population, it is called *white flight*. When they less often move into multi-ethnic neighbourhoods, it is called *white avoidance*. *Ethnic attraction* is when ethnic minorities more often move into such neighbourhoods, and *ethnic retention* is when they tend to remain and less often move away from these neighbourhoods.

Several studies in the US and in Western Europe have found evidence of selective migration among the native majority connected to the ethnic composition of neighbourhoods. The European studies indicate that white flight is less important than white avoidance. A Danish study concludes that white avoidance has been the far most important reason for the creation of neighbourhoods dominated by ethnic minorities. The study also showed that ethnic retention did not have importance. Ethnic minorities did not remain a longer time in multi-ethnic neighbourhoods than in other places. Furthermore, it was shown that white avoidance is connected to income. The income of in-moving natives sharply decreases with an increasing proportion of ethnic minorities in the neighbourhood, leading to natives in multi-ethnic neighbourhoods mostly having quite low incomes.

Reasons for white avoidance might be very different from flight. In the case of flight, the movers have themselves experienced how it is to live in such a neighbourhood and have a specific knowledge about the living conditions in them. Contrary to this, white avoidance is based on people's imaginations of how it is to live in multi-ethnic neighbourhoods and on a publicly dispersed reputation of these neighbourhoods, often communicated by the media in a distorted way.

Selective migration among the native majority is somewhat connected to preferences for avoiding neighbourhoods dominated by ethnic minorities or to preferences for staying in a homogeneous neighbourhood dominated by the majority in the country. It has been shown that such preferences tend to be more pronounced among socio-economically weak and vulnerable populations and among individuals with conservative political views. Respondents with better education and income less often have such preferences, while unemployment status increases them. Also among older people preferences are higher.

It could be assumed that preferences for living in ethnically homogenous neighbourhoods are influenced by the extent to which natives in general are exposed to ethnic minorities. There is evidence that resistance against living with ethnic minorities is less significant among residents who actually live close to them in multi-ethnic neighbourhoods. On the other hand, it has been shown that resistance against mixed neighbourhoods increases with the size of the immigrant population in the actual city. This is in accordance with what has been found in the so-called *contextual contact literature*, which indicates that personal contacts decrease resentment against ethnic minorities, while more impersonal contacts, found in larger urban areas rather than small neighbourhoods, increase negative attitudes to them. Some studies have thus shown that natives living in multi-ethnic neighbourhoods are less negative to ethnic minorities and that this cannot be explained by that those natives, who are opposed to immigrants, more often have left the neighbourhoods.

Several assumptions have been formulated about the reasons for why natives in Europe and North America (in the US, white Americans) have aversions towards living in ethnically mixed neighbourhoods. An important question is if natives avoid or leave multi-ethnic neighbourhoods because they have a general antipathy against some minority ethnic groups and are averse to living amongst them. American research has found that preferences for location are influenced by prejudices about other ethnic groups. This has to some extent been confirmed by European research. Especially, it has been shown that the perception of threat from ethnic minorities, concerning competition for jobs, public expenses, cultural concerns and crime, affects resentment against ethnic minorities and preferences for ethnically homogenous neighbourhoods. It is possible, however, that preference for a homogenous neighbourhood is not as much a result of racial prejudices as of a desire for cultural homogeneity, also called ethnocentrism.

Attitudes among the white majority to ethnic minorities and preferences for ethnically homogenous neighbourhoods vary greatly across European countries (see Chapter 8). Two structural attributes, most often used as determinants of attitudes towards minority populations, are the relative size of the ethnic group population and the economic conditions that prevail in the host society. It has been found that the larger the size of the minority population and the more depressed the economic conditions are, the greater the hostility and prejudice are towards minority populations. Countries with a general high perception of threat from ethnic minorities also tend to have stronger preferences for ethnic homogeneity in neighbourhoods. The interesting thing is that a reverse connection can be demonstrated between proportion of ethnic minorities in a country and preferences for ethnic homogeneous neighbourhoods. Countries with high immigration tend to less often prefer ethnic homogeneity, and countries with fewer ethnic minorities more often prefer ethnic homogeneity. Two reasons for this can be proposed; either that more personal contacts between natives and ethnic minorities have resulted in less resentment towards multi-ethnic neighbourhoods in countries with a larger population of ethnic minorities, or that other differing conditions across countries have importance.

It has been proposed that resentment towards other ethnic groups is not only a result of ethnic prejudices but also of that people have developed a perception of different ethnic groups as having different social status. It is said that negative preferences for living in neighbourhoods dominated by other ethnic minorities are caused by a perception of these neighbourhoods as inhabited by low-status groups. However, strong evidence has not been found that perceived differences in status in general play a role in residential integration attitudes.

The most extensive discussion in the literature has been about to what extent white flight and avoidance are caused by other conditions in multi-ethnic neighbourhoods than the ethnic composition of residents. The so-called *racial proxy hypothesis* claims that selective migration of natives is caused by social problems often found in these neighbourhoods, such as antisocial behaviour and crime, problematic schools, low quality housing and deteriorated buildings. Multi-ethnic neighbourhoods often have a bad reputation and a low social status. It has been shown that the ethnic composition of neighbourhoods has a significant negative effect on their reputation and status, and that residents in the areas are conscious about this. It has also been proposed that multi-ethnic neighbourhoods are less attractive because ethnic diversity in communities undermines trust among people. There are mixed conclusions about the racial proxy hypothesis in the literature. Some of the studies have been based on quantitative data on mobility to and from neighbourhoods, including data on the social composition of the neighbourhoods. These studies show that social variables have importance, but not a decisive one. This was also found in the Danish study described in Chapter 5. Other studies, based on surveys among residents, look at the connection between wishes to move, dissatisfaction with the neighbourhood and complaints about the ethnic composition and other conditions. Some of these studies claim that the proportion of minorities has little importance when controlled for dissatisfaction with neighbourhood problems, while others found a marked influence of the ethnic composition.

A growing number of studies have looked at the complex interplay between schools and neighbourhoods. They all find some kind of connection between school problems in ethnic neighbourhoods and neighbourhood dissatisfaction or wishes to move away, but not always very strong. It has been found that natives have a different perception of the importance of the ethnic composition in schools than minorities and that this affects differences in their neighbourhood dissatisfaction and wishes to move away. A study in the Nordic countries showed that the role of schools in ethnic segregation is very context-dependent and differs much across countries and cities. Moreover, the reasons for the effect of local schools on dissatisfaction are not quite clear. Many bilingual pupils in a school could result in a greater strain on resources, leading to a lower quality of education. Problems in schools can also just mirror social problems in the neighbourhood.

The importance of housing policies and housing markets (Chapter 6)

Housing policies and housing markets diverge greatly across European countries, and this is only to some extent the consequence of differences in welfare state systems. It seems that every country has followed its own course in housing policy, creating very different housing markets with different consequences for ethnic minorities. In general, the housing situation of ethnic minorities is poorer than that of the native majority.

Housing markets are important causes of ethnic segregation, as is shown in Chapters 6 and 7. If ethnic minorities are confined to certain parts of the housing market and this kind of housing is concentrated in certain parts of the cities, segregation will be strong. The most important division of housing markets is between different kinds of housing tenures. Depending on how tenures are designed, the housing market can be more or less *segmented.* Segmentation is created when different tenures to a great extent are made available and attractive for different groups of households, for example divided by income. Segmentation often means that high-income groups are concentrated in certain parts of the housing market, mostly owner-occupied detached housing, while low-income groups mostly reside in poor private rental housing or social housing.

Ethnic minorities in Europe have in general lower incomes than average. It is therefore of importance for them to what extent housing markets are providing good housing opportunities for families with lower incomes. In the research literature on comparative housing policies across countries, a distinction has been made between dual and unitary housing markets. Dual markets are characterised by a high degree of social segmentation and larger income differences between homeowners and renters. This is not only because income inequality is higher but also because owner-occupation is more difficult to obtain for families with middle and low incomes due to high property prices, difficulties in getting loans or high interest rates on mortgages.

Another distinction is made between dual and unitary rental markets. In dual rental markets there is a smaller social housing sector, often with a restricted access for only the very poor, while unitary rental markets have a stronger social housing sector that competes on more equal terms with private renting and has a more open access. Countries with both a dual market in general and with a dual rental market have the highest social segmentation with large income differences across tenures. If the rental sector in general is small, low-income groups experience increased difficulties in obtaining decent housing. However, in countries with a unitary housing market and unitary rental market, low-income groups have much better chances to find good housing, especially if there is a larger social housing sector.

Chapter 8 indicates that *income segmentation* of housing markets, regarding homeownership/renting, differs greatly across countries. The most important factor seems to be that countries with high segmentation often also have larger rental sectors. This indicates that lower income segmentation in some countries might be a result of families with lower incomes sometimes being forced into poor and overcrowded homeownership because of lack of rental housing.

Ethnic segmentation of housing markets concerns to what extent ethnic minorities are evenly or unevenly distributed across housing tenures. Data on the extent to which ethnic minorities are under-represented in homeownership compared to the native majority show that this differs greatly across countries. Ethnic segmentation regarding renting/homeownership tends to be smaller in countries with a high homeownership rate. As for income segmentation, this can be explained by that ethnic minorities sometimes are forced into homeownership, often overcrowded and of low quality, because of lack of rental housing. Chapter 8 demonstrates that ethnic segregation tends to be more frequent in countries with a high ethnic segmentation of housing markets.

Another explanation for differences in homeownership rates could be that ethnic minorities have higher incomes in some countries compared to others. Especially if their incomes compared to the national level are higher, one should expect a higher homeownership rate. However, the data analysed in Chapter 8 show that there is not a simple connection between income levels of ethnic minorities in countries and ethnic segmentation of housing markets.

The most fundamental assumption is that ethnic segmentation is connected to income segmentation of housing markets. The analyses in Chapters 6 and 8 show, however, that there is not a connection between these two types of segmentation. Ethnic segmentation thus cannot be explained by income segmentation. Besides differences in immigration to the countries, two factors might be the reason for this. The first is that the character and size of rental markets, which differ greatly across countries, have decisive importance. The other is that the housing situation of ethnic minorities cannot be explained by their lower incomes. Other explanations are important.

As discussed in Chapter 6, the main reasons for why the housing situation of ethnic minorities differs from natives can be assigned to:

- Their lower incomes
- Special housing preferences and preferences for ethnic neighbourhoods, which sometimes are dominated by certain tenures
- The character and extent of the housing supply, especially rented housing, that is accessible for ethnic minorities, and especially scope of and access to social housing
- Defective knowledge about how to act in the housing market and missing social relations to housing decision makers
- Discrimination from landlords, banks and real estate agents

One of the explanations for the defective connection between income segmentation and ethnic segmentation is that ethnic minorities have lower preferences for homeownership than natives, also when correcting for lower incomes (as discussed in Chapter 2). Moreover, preferences for living in neighbourhoods with an ethnic social network (see Chapter 3) influence what kind of tenures and dwellings ethnic minorities try to obtain and what kind of accommodation they can access.

The options for ethnic minorities in the housing market and the difficulties they meet in their efforts to get decent housing are very much dependent on the supply of housing in different housing tenures and how housing policy regulates access and housing costs. The supply of low-cost rental housing is very important for ethnic minorities, but there are also some other properties of the housing market that have importance.

The supply of social housing and rules for access vary greatly across countries. Sometimes social housing is open for all house hunters and sometimes access is restricted to families with low incomes or specific needs. There is usually a connection between extent of supply and rules for access where countries with small social housing sectors also have strong restrictions on access. Large social housing sectors with free access are the most favourable for ethnic minorities. In some cases, however, restricted access can be an advantage for them if the supply of housing is sufficient to meet their demand, but this can result in a strong ethnic segmentation of the housing market, as is shown for Finland in Chapter 6. It can also be an advantage for ethnic minorities if local authorities can dispose of some of the vacant dwellings in social housing. In some countries this option has been used to house vulnerable ethnic minorities and especially newly arrived refugees.

Defective knowledge about how to act in the housing market and missing social contacts to decision makers in the market have also been detected as limitations for ethnic minorities' housing options (Chapter 6). This especially concerns private renting, where good contacts to landlords are important, but also in some cases social housing and cooperatives. Lack of transparency in allocation procedures and intricate rules make it more difficult for ethnic minorities to find housing.

Discrimination in the housing market can make it more difficult for ethnic minorities to get access to adequate housing. There is strong evidence of discriminatory practices in the housing market, which harms ethnic minorities. In particular private, and to a certain extent social landlords, exclude ethnic minorities from their housing. Moreover, discrimination from banks and real estate agents might be one of the reasons for lower homeownership rates. Chapter 8 indicates that discrimination in the housing market varies a lot across ethnic groups and countries. On average North Africans seems to be the most discriminated against group, followed by Sub Saharan Africans and Turks, but the differences can to some extent be explained by differences across the countries where they live.

The occurrence of discrimination in social housing depends on to what extent rules for access are transparent and give equal opportunities for all independent of ethnicity. In some cases 'hidden' maximum quotas, for example for house hunters without employment, is in praxis used to limit and restrict the concentration of immigrant families on housing estates. In other cases, unclear rules provide the basis for local decisions that make room for discrimination.

In private renting there are stronger bases for discrimination if supply is small compared to demand. This appears when the supply of rented housing is too small or if rents are below market prices because of rent control. In the case of excess demand, landlords may have stronger incentives to choose new tenants from their own network or from those who belong to the native majority.

The importance of urban policies and the spatial distribution of housing (Chapter 7)

The literature on causes of spatial segregation has mostly focused on the effects of social and cultural differentiation of the population in cities. Spatial segregation has mainly been understood as a direct consequence of how the population has been divided into separate social and cultural groups competing for attractive neighbourhoods, resulting in a spatial sorting of groups into different parts of the cities, separated from each other. Lately it has been pointed out, however, that there are more differentiated and context-bound explanations of segregation, where spatial characteristics of cities are included as important factors. One of these characteristics is the spatial housing structure, especially the segmentation of housing markets and the spatial separation of different housing tenures.

The distribution of people in space is created by interaction between social and cultural differentiation among residents and *spatial differentiation* of cities. This spatial differentiation is a product of the social, physical and functional structure of the city, a structure that is continuously changing due to economic investments and disinvestments as people and functions are redistributed in space. Cities become divided into identifiable areas that can be relatively homogeneous, but exhibit distinctive characteristics that are very different from other neighbourhoods. Such areas differ not only by virtue of different physical qualities, such as housing stock, architecture, natural attractions, and access to transport facilities and service supply, but also by less tangible qualities such as social and cultural image, status and security. Preferences for living in various neighbourhoods can vary across households with different needs and life-styles, but there will always be some common values among people that result in some neighbourhoods being more appreciated by the majority, while other neighbourhoods are seen as very unattractive. The greater differences found between the attractiveness of neighbourhoods in a city, the larger the *spatial inequality* will be. Segregation is not a simple consequence of social and ethnic inequality, but is also affected by spatial differentiation leading to spatial inequality. Segregation and increasing spatial inequality are mutually self-perpetuating processes because the social status and cultural identity of individual urban areas are determined by the composition of residents. Spatial differentiation leads to segregation, while segregation creates spatial differences and inequality.

Different kinds of housing are to a different extent attractive and accessible for different social groups. The spatial distribution of housing is thus an important factor in creating spatial inequality. It could be argued that this spatial distribution is just a mirror of housing demand from different groups, whose preferences for location create the spatial distribution of housing tenures. Housing structures of cities have, however, been created over a long time. This especially concerns the central cities where old and sometimes dilapidated and obsolete housing remain. The spatial distribution of housing is, moreover, sometimes highly regulated by national housing policies and urban planning, which interfere with the natural processes of market segregation. It has been found that differences in housing policy, and the resulting housing markets in the countries, together with planning systems, have played a very important role in explaining differences in ethnic segregation across countries.

The degree of segregation and the character of ethnic neighbourhoods vary greatly across European countries. The settlement pattern of ethnic minorities has to some extent been determined by formation of neighbourhoods dominated by ethnic minorities where processes of selective migration have made it easier for immigrants to obtain housing and where a tight ethnic network has been established that has been attractive for immigrants. There is a mutual relationship between the character of ethnic concentration, selective migration and preferences for ethnic enclaves. The existence and intensity of ethnic neighbourhoods have depended on the size and composition of ethnic groups and the extent of ethnic segregation in countries and cities. The formation of ethnic neighbourhoods has depended on to what extent local neighbourhoods in the first place have been available with accessible housing for immigrants. Formation of ethnic neighbourhoods has thus depended much on the historic spatial structure of urban housing markets.

The structure of cities has been changed by development of new housing estates on the urban fringe, but also by urban renewal of older housing and redevelopment of converted land. These changes have to some extent been influenced by urban renewal programmes and by public planning and land control. Urban renewal in Europe after the World War was directed towards problems of decay and obsolete housing in the older and central parts of cities. It was a policy that often led to a displacement of the residents from older neighbourhoods in the central cities to the suburbs, which also affected ethnic minorities. The effect of urban renewal for the location of ethnic minorities has, however, depended much on the extent of renewal and the kind of policies that have been used. In some cases there has been more focus on giving residents options for staying in the renewed neighbourhoods; in others gentrification has been the main objective.

The spatial structure of housing markets is not only dependent on public planning and intervention in the use of land but also on the character of institutions and markets that shape and change the urban structure. It is especially important whether the market is dominated by large builders and developers or by small-scale builders. Large firms and developers tend to build larger and more uniform housing areas, especially if they are operating in cities where the local authorities do not own the land and do not regulate land use much. It is therefore decisive how planning and housing systems are combined. High ethnic segregation can thus either be a result of a combined liberal housing and planning system or of a strongly segmented housing system combined with a planning system that separates tenures in space. A lower rate of segregation can be obtained if housing policies result in a low degree of ethnic segmentation of tenures or if a strong planning system results in spatial dispersal of segmented tenures. It has thus been shown that there is not a straightforward connection between strong planning and control with land use and less segregation. Sometimes strong planning has resulted in high separation of housing tenures and strong segregation.

It has been shown that an uneven spatial distribution of different housing tenures has quite a strong influence on segregation. The analyses from four Nordic capitals, described in Chapter 7, revealed a strong connection between tenure composition of neighbourhoods and ethnic concentration in all four cities, which illustrated the importance of the housing market for the spatial distribution of immigrants and for their separation from natives. It was also shown that the general degree of ethnic segmentation of housing markets did not seem to be the most important factor for segregation. The spatial distribution of housing tenures is more important. The city of Helsinki in Finland is a specific example of how strong planning objectives of mixing housing tenures results in less ethnic segregation even if the city has a very segmented housing market.

Some earlier research focused extensively on the importance of the size and character of the social housing sector in the cities as a major explanation of differences in segregation. This could be ascribed to the existence of large spatial concentrations of social housing with many immigrants. It was assumed that a marginalised social sector with high concentrations of low-income groups also will result in high ethnic segregation. It has been shown, however, that ethnic segregation in Northern and central Europe not only depends on spatial concentrations of social housing but that also other tenures can provide a basis for concentration of ethnic minorities.

Why does ethnic segregation differ so much across countries? The importance of welfare state systems (Chapter 8)

In Chapter 8 comparisons are performed across European countries to elucidate differences in ethnic segregation and to compare some of the conditions in the countries, which we in the other main chapters have identified as crucial for ethnic segregation. This concerns preferences among the whole populations for not living close to ethnic minorities and how this is influenced by negative attitudes towards ethnic minorities. It also relates to differences in housing markets and their social and ethnic segmentation and how these conditions affect ethnic segregation and the housing situation of ethnic minorities. Finally, we discuss the importance of differences in planning systems across the countries.

A main subject is to examine how differences between welfare systems affect ethnic segregation. West European countries have in the literature been divided into different kinds of welfare state systems. Mostly they have been divided into four groups:

1. Social Democratic welfare cluster (Sweden Denmark, Finland, Norway, Netherlands),
2. Corporatist welfare cluster (Germany Austria, France, Belgium),
3. Liberal welfare cluster (UK, Ireland, Switzerland, Luxembourg and to some extent also Belgium),
4. Latin Rim welfare cluster (Greece, Portugal, Spain, Italy).

We have, moreover, in Chapter 8 designated a fifth group of the newer EU members from Eastern Europe.

Welfare regimes can, in multiple ways, affect and shape the patterns of segregation. Income redistribution, access to citizenship, education and health care, etc. are all welfare factors that influence immigrants' chances of having success in their new country and their ability to integrate and to choose by themselves where they want to settle. Housing systems, especially the extent of social housing, affect the housing options of ethnic minorities, and the location of different kinds of housing often determines where they can settle.

Social Democratic welfare states have universalistic systems that promote an equality of high standards, rather than an equality of minimal needs. This implies publicly supplied welfare services, to reduce the social division introduced by market-based access to welfare services, as well as pre-emptively socialising the costs of caring for children, the aged and the helpless, instead of than waiting until the family's capacity to support them is depleted.

Liberal welfare states are characterised by modest, means-tested assistance, and targeted at low-income recipients. Their strict entitlement rules are often associated with stigma. This type of welfare state encourages market solutions to social problems – either passively, by guaranteeing only a minimum, or actively by directly subsidising private welfare schemes.

Corporatist and Latin Rim welfare systems are typically shaped by traditional family values, and tend to encourage family-based assistance dynamics. Social insurance in this model typically excludes non-working wives. State assistance will typically only step in when the family's capacity to aid its members is exhausted. In the Latin Rim systems, income polarisation is counterbalanced by Christian social policy (family, patrimony) and informal access to employment and resources.

As ethnic minorities mostly have lower incomes than average, their situation is influenced by the general degree of social inequality. Income inequality varies significantly across the countries and welfare clusters. In Western Europe, inequality tends to be highest in the Latin Rim countries followed by the Liberal cluster and the Corporatist cluster. Equality is, as could be expected, highest in the Social Democratic cluster. Among the Eastern countries the situation is more diverse.

Ethnic segregation is influenced by the size and composition of ethnic minorities, which diverge greatly from country to country. In former colonial power countries, immigrants from their colonies are often dominating; in others the character of labour import and refugees has had essential importance. The proportion of foreign born varies a lot across countries, also within welfare clusters. The highest proportion of foreign born is found in the Liberal and the Corporatist clusters. The Social Democratic have had lower immigration, except for Sweden. In the eastern part of the EU, most countries have fewer ethnic minorities and they probably consist mainly of immigrants from Eastern Europe, Russia, Ukraine, etc.

The ability of ethnic minorities to choose the housing they want and to decide where to live in the city depends on their economic resources and to what extent their income is comparable with the native majority. The income of ethnic minorities varies from country to country in pace with the general income level, but not quite in a consistent way. The variation across countries in incomes for Non-EU citizens is not a simple consequence of different national income levels, as the relationship between the incomes of ethnic minorities and the average for the country varies greatly. In some countries their incomes are quite close to the national average; in others ethnic minorities are much poorer than is normal for the country. This variation cannot be assigned to welfare state systems, as it is difficult to uncover major differences across welfare state clusters. Ethnic minorities in the Latin Rim countries are the ones who have the lowest incomes compared to the national average, but the difference from other clusters is not great.

The extent of ethnic segregation across countries

It has been found that levels of spatial segregation of single groups differ greatly across welfare clusters. Cities characterised by a liberal welfare regime have the highest levels of segregation for some special groups, especially in the city centre areas. At the other extreme, cities characterised by a Corporatist welfare regime, followed by cities with a Latin Rim regime, scored the lowest levels of group segregation, and the areas of ethnic over-representation were located either in the inner city or in the inner suburbs. Cities within the Social Democratic cluster were an intermediate case, with the highest group concentrations outside the city centre.

The extent of spatial concentration of ethnic minorities in general has been measured by data from the European Social Survey where people were asked about the ethnic composition of their neighbourhood. This data indicates that ethnic concentration varies greatly across countries, but that there also seem to be some systematic differences across welfare state clusters. Corporatist welfare states seem to have a relatively high concentration of ethnic minorities, especially in France, but also Liberal states like the UK and Switzerland have it. Scandinavian countries and to some extent Holland are at a lower level of ethnic concentration. Southern countries like Spain and Greece are measured to have a medium concentration.

The data also reveals to what extent the native majority resides in ethnic homogeneous neighbourhoods with very few ethnic minorities. Also here there are deviations across countries, but not quite so significant differences across welfare state clusters, except for the Eastern countries and to some extent the Social Democratic cluster where it is a little more common to live in homogenous neighbourhoods.

Everything else being equal, it must be expected that the probability of getting high concentrations of ethnic minorities depends on how large a population of ethnic minorities are found in a country. A comparison across countries shows that there, as should be expected, is a straightforward connection between ethnic concentration and the proportion of foreign-born residents. However, some countries have much more ethnic concentration than should be expected from their proportion of ethnic minorities. France, the Netherlands and Spain are countries that seem to be more segregated compared to the proportion of foreign-born residents in the countries. This can to some extent be explained by that they immigrated earlier and have more descendants of immigrants that are perceived as ethnic minorities in the survey. Countries with relatively less segregation compared to the proportion of foreign born are found in all welfare clusters, but three of them belong to the Social Democratic cluster (Norway, Sweden and Finland), besides Germany and Ireland.

Preferences for ethnic composition of neighbourhoods

As discussed in Chapter 5, preferences for the ethnic composition of the neighbourhood among the ethnic majority, and the extent to which they avoid moving to multi-ethnic neighbourhoods, is one of the major factors explaining ethnic segregation. Data from the 2002 European Social survey (ESS) indicates that preferences for living in ethnically homogeneous neighbourhoods, with few ethnic minorities, vary greatly both across countries and inside welfare clusters. Preferences fluctuate from 45 per cent in Greece to 18 per cent in Switzerland. It is difficult to find systematic differences between welfare clusters. In the Social Democratic cluster, Sweden and Norway have the weakest preferences; in the Corporatist cluster it is Germany and France, and in the Liberal cluster the UK and Switzerland. All the countries in the Latin Rim and Eastern clusters, except for Italy, have higher preferences.

The inclination among the native majority to avoid neighbourhoods with ethnic minorities can be a result of prejudices towards them. Data from a study using the ESS 2002 has been used to illuminate differences across countries concerning anti-minority attitudes in the populations. This again shows pronounced differences across the countries in the extent of negative attitudes towards ethnic minorities. However, there is here some connection to the type of welfare state. Negative attitudes are lowest in the Nordic Social Democratic countries. They are also low in Liberal states such as Luxembourg and Switzerland, which can be explained by that these countries mostly have received immigrants from Western countries. Stronger anti-minority attitudes are found in a mix of countries belonging to the Liberal welfare cluster (UK and Belgium), the Latin Rim welfare cluster (Greece, Portugal) and some East European countries. Countries from the Corporatist welfare cluster (Germany Austria, France) have in general medium strong negative attitudes.

A connection is demonstrated between preferences for ethnic homogeneity and negative attitudes. Countries with stronger negativity also tend to have stronger preferences for ethnic homogeneity in neighbourhoods. To the extent that these preferences lead to white avoidance, it can thus be concluded that ethnic segregation is influenced by political and economic conditions in countries, which leads to negative attitudes towards ethnic minorities among the native majority.

It should be assumed that countries with higher preferences for ethnic homogeneous neighbourhoods also would have stronger ethnic segregation and concentration. However, this does not seem quite to be the case. There is not found a systematic increase in ethnic concentration with increasing preferences for homogeneity across countries. There is also not a straightforward connection between preferences for homogeneity and the proportion of residents actually living in such neighbourhoods.

The somewhat curious conclusion from this is that even if it is shown that negative attitudes towards ethnic minorities affect preferences for not mixing with ethnic minorities, these negative attitudes do not seem to have consequences for the extent to which the actual residence of the native majority is separated from ethnic minorities. The explanation might be that differences across countries concerning residential preferences among ethnic minorities and the native majority do not constitute a decisive determination of ethnic segregation. As described in Chapters 6 and 7, ethnic segregation depends very much on the residential options for both ethnic minorities and natives constituted by housing markets and structural properties of cities.

Housing markets across countries and their importance for segregation

Housing opportunities are decisive for where ethnic minorities can settle in the city. If they are confined to limited parts of the housing stock, and this housing is located in specific parts of the city, the probability for ethnic segregation is high. It has been shown that ethnic minorities are very dependent on rented housing, mostly social housing or housing of low quality.

As described in Chapter 6, housing policies vary extensively across countries and for different reasons; there is not always a straightforward connection between housing policies and welfare policies in general. Some countries with strong welfare have chosen more liberal housing models, while some countries with weak welfare have implemented a more social housing policy.

The supply of low cost rental housing, especially social housing, and the rules for access to this housing, are very important for ethnic minorities. Countries in the Latin Rim cluster and in Eastern Europe nearby generally have a low proportion of rented housing and a small stock of social housing. Some of the southern countries also have up to 20 per cent private renting. The largest non-ownership sectors are found in the Social Democratic and Corporatist clusters. Social housing is mostly found in Social Democratic countries, but is also important in some other countries like Austria, France and the UK. Social housing is nearly non-existent in many countries, especially in Southern and Eastern Europe. Moreover, the role and organisation of social housing varies in important ways across welfare clusters. In the Social Democratic cluster, several countries have a so-called *Universal allocation model* that makes social housing available to the whole population regardless of their income. Many countries in the Corporatist and Latin Rim clusters have used what have been called *Generalist models*. Social housing in these countries is aimed only at those households for whom the market is unable to deliver housing of decent quality at an affordable price. The remaining countries use what has been called the *Residual allocation model*, which aims to allocate housing to a very restricted category of beneficiaries, usually very vulnerable households that are heavily dependent on state benefits.

The housing options of ethnic minorities are expected to be influenced by the options for low-income groups in general, because they mostly have lower incomes. The extent to which there is a social division, called income segmentation of housing markets, affects the housing options of ethnic minorities. Chapter 8 demonstrates that income segmentation of housing markets varies greatly across countries as well as within welfare clusters. The five countries with the highest income segmentation are spread across all four clusters. The most important factor seems to be that countries with high segmentation often also have larger rental sectors. This indicates that lower income segmentation in some countries might be a result of families with lower incomes being forced into homeownership, often of low quality, because of lack of rental housing.

Chapter 8 indicates that ethnic spatial concentration depends on ethnic segmentation of housing markets. Living in multi-ethnic neighbourhoods is more frequent in countries where ethnic minorities are much under-represented in homeownership. Ethnic segmentation varies a lot inside welfare clusters, but it tends to be lowest in the Social Democratic cluster, a little higher in Corporatist states and highest in the Liberal and Latin Rim clusters. Some of the countries with the largest difference between natives and ethnic minorities are found in Liberal states and Latin Rim States, but also in some Corporatist countries and Social Democratic countries. These differences can only to a limited extent be explained by the fact that the incomes of ethnic minorities in some of the countries are closer to the national average.

The data indicates that there is not a simple connection between income segmentation and ethnic segmentation, but that there seems to be a reverse relationship between them. The three countries with the highest ethnic segmentation (Italy, Austria and Greece) also have quite low income segmentation. The UK has the highest income segmentation, but only a mediate ethnic segmentation. This data insinuates that there is not a simple connection between welfare state systems, income levels and ethnic segmentation of housing markets. The extent and character of immigration might have had some influence on the differences, as some countries have received more immigrants with low education or refugees with social problems. Most importance must, however, be attached to differences in housing policies and housing markets across welfare clusters that provide different opportunities for low-income groups and for ethnic minorities. It should be assumed that if homeownership is more difficult to access for low-income groups, it will also be more difficult for ethnic minorities because they in general have lower incomes.

Among the important conditions for housing options of ethnic minorities is discrimination in the housing market. The findings indicate that discrimination varies a lot across ethnic groups and countries. On average North Africans seems to be the most discriminated against group, followed by Sub Saharan Africans and Turks, but the differences can to some extent be explained by differences across the countries where they live. Discrimination is mostly found in Latin Rim countries, while it is lowest in the Social Democratic welfare states and in the UK. The Corporatist states, Belgian, Germany, Austria and France, all have quite high degrees of discrimination.

One of the signs of difficulties for ethnic minorities to find decent housing is the degree to which they live in overcrowded dwellings. Overcrowding among ethnic minorities is especially often found in Latin Rim and Eastern countries. In average for the EU, Non-EU citizens live about 80 per cent more often in overcrowded housing. Compared to the housing situation of the native majority, the largest *relative housing deprivation* among Non-EU residents seems most often to be found in countries belonging to the Liberal and the Corporatist countries, the smallest in Eastern Europe and some countries in the Social Democratic cluster.

Differences in planning and building systems

As discussed in Chapter 7, both the spatial structure of cities and ethnic and social segregation are influenced by the way public authorities intervene in the development of cities by controlling the use of land. Of special importance is to what extent different kinds of housing is distributed across neighbourhoods.

Spatial planning systems are deeply embedded in the socio-economic, political and cultural context in the countries and seem to be connected to welfare state systems. In the Social Democratic welfare cluster of countries, and also in some Corporatist welfare states like Germany and Austria, there have been planning systems where public authorities are intervening heavily in the development of cities. In other countries, mostly found in the Liberal Welfare cluster, a kind of regulation has been used that has a much narrower scope with the limited purpose of regulating land use change and with an urban development created mostly by market forces. In other countries, from the Corporatist and Latin Rim clusters, planning has mostly been steered by regional objectives of economic growth and development, where local objectives of developing cities with less spatial inequality have not been much in focus. Finally most Latin Rim countries have had a spatial planning style, which takes place on the local level through building regulations. Public planning is weak and seldom interferes with the use of land.

It has been shown (see Chapters 7 and 8) that there is not a systematic connection between strong planning systems and low segregation. The Nordic Social Democratic welfare states (and the Netherlands) have, despite lower housing segmentation, a relatively high degree of spatial segregation. This is explained by the mode of building production in the countries, where building activities are dominated by large-scale builders and developers. In spite of strong planning systems and much public land ownership, the structure of the building and housing sector with large builders, developers and owners has promoted the development of large urban areas with homogeneous tenures and building types, which has promoted a spatial segregation of tenures. Also, Liberal welfare regimes tend to have large builders/developers relying more on speculative development gains than on building profits.

The Corporatist welfare state cluster has been characterised by having a more fragmented building industry than in both former cases. The Latin Rim welfare states of the Mediterranean should have even smaller builders than their Corporatist counterparts. In both Corporatist and Latin Rim regimes, a division is found between different kinds of firms that perform housing production for respectively high, middle and low-income households, and are producing at a relatively small scale.

How do welfare state systems influence ethnic segregation?

Welfare states are characterised by higher incomes and wealth and less inequality, because of more equal wages and better social benefits. They have more public support for housing and have better planning instruments that can be used to manage the spatial development of cities. One should expect that lower and more unequal incomes also would result in stronger social segregation in countries, which might lead to ethnic segregation. Moreover, good housing opportunities for the poor should result in opportunities for ethnic minorities and more options for where to live in a city. Strong planning systems might be used to avoid a concentration of housing for low-income groups and ethnic minorities in specific neighbourhoods.

It has been shown, however, that the incomes of ethnic minorities, compared with the national average, do not vary systematically across clusters of welfare states, which means that general income inequality does not lead to relatively lower incomes among ethnic minorities. Moreover, a comparison of the Gini index for the countries with the indicator of ethnic concentration does not show a systematic connection between these two measures. Income inequality thus does not automatically lead to more ethnic segregation and differences in social inequality across countries thus cannot explain variation in ethnic segregation.

It is, however, shown that spatial concentration of ethnic minorities to some extent is connected to the size of the ethnic minority population in the countries. As expected, the presence of neighbourhoods with many ethnic minorities will, everything else being equal, be more common in countries with a larger minority population.

Segregation is influenced by the preferences for location among the native majority. Preferences for living in ethnic homogeneous neighbourhoods vary greatly across countries and also inside welfare clusters. These preferences are stronger in countries where the population shows more negative attitudes towards ethnic minorities, and such attitudes are more often found in countries with less extensive welfare systems. However, there is not a significant connection between these preferences and the extent to which the actual residence of the native majority is separated from ethnic minorities. The explanation might be that differences concerning residential preferences among ethnic minorities and the native majority across countries do not constitute a decisive determination of ethnic segregation.

Housing conditions and housing options are better for ethnic minorities in the more developed welfare states. The ethnic and social division of housing markets, segmentation, is lower. This is a result of more rental dwellings, and especially social housing, and less discrimination. Ethnic minorities have better opportunities in these markets to choose where to live in the city, which should lead to less segregation. This has not always been the case as the most favourable kind of housing, social housing, often has been concentrated in specific neighbourhoods in the cities, where strong concentrations of ethnic minorities have appeared. In the least developed welfare states, ethnic minorities often are more dispersed in the housing stock, and the location of ethnic minorities more dependent on market processes of social and ethnic segregation, which especially are found in Liberal welfare states. Minorities have often been confined to low quality housing in private renting or homeownership concentrated in more or less declined neighbourhoods.

2 Housing preferences and choices of ethnic minorities

Co-authored with Timo Kauppinen[1]

Ethnic segregation is influenced by individual preferences among ethnic minorities (and the native majority) for housing and location. In this chapter is examined if ethnic minorities have different preferences for housing than the native majority, which might have importance for their choice of residence.

Evidence shows that the housing situation of ethnic minorities in most countries diverges much from that of the native population. These differences cannot be fully explained by lower incomes and education among immigrants (Flippen 2001; Alba and Logan 1992; Sinning 2010;Kauppinen et al. 2015). They could be a consequence of restricted options for ethnic minorities in the housing market, as will be discussed in Chapter 6, but it could also be that some ethnic minorities have housing preferences that differ from those of the native majority.

Housing outcomes of immigrants and other ethnic minorities can to a great extent be expected to have the same explanations as those for the majority ethnic groups. That is, they depend on family situation, economic resources and local housing market possibilities. The family situation, family-related events and changes in the life cycle have been shown to be important determinants for housing outcomes (e.g., Clark and Onaka 1983; Howell and Frese 1983; Clark et al. 1997; Feijten and Mulder 2002; Mulder 2007; Skifter Andersen 2011).

In this chapter we will discuss the possible deviations between the housing preferences and choices of ethnic minorities and the preferences and choices of the native majority in Western countries, as it is known from the research literature. Such differences can be assigned to deviations among ethnic minorities concerning family structure and family careers, housing norms and economic behaviour, which influence what kind of housing is preferred. Of importance is, for example, if extended families are typical and when and how young people are when leaving their parents. Of particular importance is the extent to which ethnic minorities prefer to rent or to own their dwelling and why. This question has been in focus for comprehensive research. The literature on this will be discussed in detail.

In the last part of the chapter the results from a study in three Nordic capitals will be used to illuminate the propensity of ethnic minorities to move into homeownership and the factors that influence this. Despite differences in housing markets and composition of ethnic minorities across these three cities, a very similar pattern is found. In all three cities ethnic minorities have a much lower tendency to move into homeownership, and this cannot be explained by differences in socio-economic background factors.

Studies of housing preferences among ethnic minorities

Studies of residential preferences have been based on measurement of either: 1. Stated preferences, or 2. Revealed preferences (Floor and Van Kempen 1997; Timmermans et al. 1994; Coolen and Hoekstra 2001). Stated preferences are found by asking people directly about how they would prefer to live and why. Revealed preferences are found by examining how people actually live. It can be either among recently moved households or all.

Both methods have methodological problems. By the method of revealed preferences, the problem is that the actual housing situation does not necessarily reflect the underlying preferences. This is because, first, the real opportunities on the housing market and the individual's financial resources play a role. Even when trying to identify the housing options and economically conditioned preferences, the actual situation can give a false picture, because regulations of the market and imbalances may lead to one being unable to obtain the optimal housing situation (Timmermans et al. 1994; Skifter Andersen 2011) Especially for ethnic minorities this is important, as they for different reasons have fewer options, as is described in Chapter 6. In addition, different circumstances impeding mobility can result in that many families, despite significant changes in housing needs, remain in homes that in the course of time increasingly come to differ from their preferences (Skifter Andersen and Bonke 1980).

By stated preferences people do not necessarily take account of the economic possibilities for realising them. This can mean that there could be significant variations between responses from households that should have the same preferences, as some have a realistic picture of their chances in the housing market – and have included this in their statements – while others have not, and therefore state preferences that are not realistic. It is a common situation in most countries that more households have preferences for homeownership than the number who can actually obtain it (Dieleman et al. 1998). Most studies on housing preferences are based on revealed preferences, where the connection between different factors and actual housing choice is determined (Dieleman et al. 1998). In these studies, the actual motives for the choice can only be assumed, based on an interpretation of the revealed connections.

Preferences for housing conditions

In sociological or socio-economic housing research, weight has been placed on socio-economic factors. Differences in family situation and changes in the life cycle are traditionally seen as the factors of greatest importance for housing needs and preferences (Skifter Andersen 2011; Clark and Onaka 1983; Howell and Frese 1983; Floor and Van Kempen 1997). Especially the changes, when children arrive to, or leave, the family, create dramatic changes in housing demand, particularly in the phase where the families establish their more permanent housing situation. The need for space, facilities and environment changes dramatically when they have children. Later, when children move away from home, space requirements are reduced, but this does not always lead to moves to another home – because people find other uses for the space and partly because household mobility decreases dramatically over the years (Rossi 1955; Skifter Andersen 2011).

While there is a wide range of studies on revealed preferences among ethnic minorities, based on quantitative data on their housing situation, studies based on stated preferences are rare. Most of them are based on qualitative data gained from personal interviews (Bowes et al. 1997; Phillips 2009; Skovgaard Nielsen et al. 2015; Skovgaard Nielsen 2017; Søholt 2014).

In general, it could be expected that many immigrants are coming from countries where housing standards are low, which could affect their expectations regarding housing in their new country in Europe or North America. On the other hand, family structures in the country of origin often are very different from what is found in the new country. In many countries people live in extended families with several generations with grandparents and sometimes also other branches of the family. In a Danish survey among Non-Western ethnic minorities (Skifter Andersen 2006), 20 per cent of the respondents expressed a wish to live in extended families. In the study was shown a negative connection between indicators of social integration and the wish to live in extended families. Those who did not speak Danish properly, were without jobs or did not have contact with Danes were much more prone to prefer extended families. Also among those who prefer to live in nuclear families, some have to share their dwellings with other family members who are not able to find proper housing, because of the strong obligations to help the family which are found in many cultures, called reciprocity (Portes and Sensenbrenner 1993). Reciprocity includes the possibility of moving in with others when in need of a place to stay, thus avoiding sleeping rough in the streets. This is more important in some cultures than in others. A study in four Nordic countries (Skovgaard Nielsen et al. 2015) showed this to be very important among Somali immigrants.

In a qualitative study among Pakistanis in England (Bowes et al. 1997), most of the households at some time had other members of the family staying there. In seven instances, it was a brother or sister while, in four cases, older sons had married and a daughter-in-law had joined the household. Only one household had grandparents living in the house, while, in three cases, friends had lodged with the household. Household size and content changed over time as the needs of the basic family changed, for instance by child births or marriage among children, leading to new housing problems in the family. For the Pakistanis all this spawned a need for larger and flexible dwellings. An example of the consequences of extended families is that a family with old and handicapped grandparents had to find a dwelling on the ground floor. If new demands could not be fulfilled in the present dwelling, the families often tried to find another home. Phillips (2009) demonstrated in another British study that extended families in some cases chose to split up and live in two dwellings, but continued to function as one family. This leads to a demand for finding dwellings close to another.

Of a special importance is if grown-up children stay put in their parents' home or leave to form their own household. This influences both the kind of housing needed by the parents and for housing for the moving young people. Important are also the reasons for moving; is it to form a new family by marriage or is it to live as a single person? Research on this is sparse and points to different results for different ethnic groups. The tendency for ethnic minorities to live in extended families should imply that young people would stay longer with their parents. Zorlu and Mulder (2011) indicated, however, that Moroccans and Turks in the Netherlands tend to leave home earlier than Dutch youth. The reasons for leaving also vary between groups. Skovgaard Nielsen (2015 and 2016) found that Turks in Denmark more often than Danes left in connection with marriage. This could, however, almost be explained by socio-demographic differences. Somalis had a pattern quite similar to Danes despite lower incomes. She also examined a hypothesis that the patterns of home leaving were influenced by length of stay in Denmark by comparing descendants with those who were not born in the country. She found (like Zorlu and Mulder) that socio-demographic variables were much more important reasons explaining home leaving than being born in the country. The differences between Turks and Somalis indicate, however, that cultural differences exist between different ethnic groups.

Preferences for homeownership

The options of ethnic minorities in the housing market are, as is discussed in Chapter 6, very much determined by their potentials for purchasing their own home. Homeownership is a spatially much more accessible kind of housing than renting and it is therefore important for one's opportunities for choosing where to live in a city if one prefers homeownership and is able to obtain it. Economic resources are decisive, but it has been shown that some ethnic minority groups for different reasons more seldom move into homeownership than the native population despite having the same incomes. This is documented in detail in the last part of this chapter based on a study in the Nordic countries.

The tendency for ethnic minorities to choose rental housing is not necessarily because they prefer this tenure. In a Danish survey (Skifter Andersen 2006), there was a very high ideal preference for homeownership (75 per cent) among Non-Western ethnic minorities, and most of them came from countries where homeownership is widespread. Thus, other explanations must be found. Besides lack of financial resources, some studies have pointed to the special circumstances which immigrants are subject to.

Many of them are uncertain about their future; if they should return to their home country or should stay put in their new country. This was, for example, found among Somalis in four Nordic countries (Skovgaard Nielsen et al. 2015), among Pakistani families in Norway (Søholt 2014) and among African immigrants in Toronto (Owusu 1998). One of the consequences of this is that some immigrants prefer to invest their funds in property in their country of origin instead of homeownership. This has been found among Pakistanis in Norway (Søholt 2014) and

Somalis in Denmark (Skovgaard Nielsen 2017). Moreover, immigrants also tend to be the most vulnerable employees in the labour market, often with uncertain and short-term appointments. It is known from the housing literature that the purchase of a home is very dependent on if homebuyers have long-term expectations on an economically stable future (Skifter Andersen 2011) and a wish to stay in the same place. Thus uncertainty about the future might reduce the propensity of ethnic minorities to invest in housing (Skovgaard Nielsen et al. 2015; Constant et al. 2009).

Moreover, many immigrants still have strong connections to their country of origin that influence the way they use their financial resources. One example of this is, as mentioned, investments in the home country instead of in homeownership; another is the practice of sending remittances to family members who have been left in the country of origin. This was found among Somalis in a study in the four Nordic countries (Skovgaard Nielsen et al. 2015). Remittances were regarded as unquestionable if relatives were in need. An employed single mother in Stockholm explained: 'You can't wish for improvement of your dwelling when you know someone in Somalia is starving'. Being part of reciprocal transnational relations means that remittances can limit possibilities to buy a home. Another obstacle to homeownership among Somalis was the ban for Muslims against paying interest and making profit on housing investments, which made it difficult for them to finance homeownership. This was also found among African immigrants in Toronto by Mensah and Williams (2014). This obstacle depends on how strong the connections are to the original native culture and religion and if breaking the ban leads to social stigma from the ethnic community; thus to what extent immigrants are assimilated in new norms in the host country or are stuck in cultural norms from their past.

Limited options for immigrants in the labour market also might influence their use of savings. It has been shown (e.g. Toussaint-Comeau 2012) that ethnic minorities much more often than the native majority are involved in small-scale business such as retail shops. Starting their own business requires capital and often loans from either banks or from family and friends. In either case it becomes more difficult to obtain further loans for purchasing a house.

The literature on homeownership among ethnic minorities, and the possible reasons for why it differs from homeownership among the native majority in countries, is extensive. Specific studies of homeownership among immigrants have mostly been made in Northern America (e.g. Alba and Logan 1992; Haan 2007; Borjas 2002; Flippen 2001; Chua and Miller 2009; DeSilva and Elmelech 2012; Krivo 1995; Painter and Yu 2010; Kim and Boyd 2009). Only a few articles in English about homeownership among immigrants in Europe have been found; one from the UK (Nygaard 2011), one from Sweden (Bråmå and Andersson 2010) and three from Germany (Constant et al. 2009; Sinning 2010; Davidov and Weick 2011). Sinning's (2010) study concludes that the gap in homeownership rates between immigrants and natives in Germany differs substantially from the corresponding pattern found in the US. Most studies are based on quantitative data on where people live and try to explain tenure choice by data on income, employment, family situation, etc. Only Constant et al. (2009) is based on survey data.

The studies have mostly found that differences in the general socio-demographic determinants of homeownership are insufficient explanations for the gaps between ethnic minorities and natives (Bråmå and Andersson 2010; Davidov and Weick 2011; Haan 2007; Kim and Boyd 2009; Krivo 1995; Painter and Yu 2010; Sinning 2010; Magnusson et al. 2014). This suggests that the effects of these determinants differ between natives and immigrants. The results from a study in three capitals in the Nordic countries of Europe, described in a section below, confirm this.

The effects of specific socio-economic and demographic factors on homeownership have also been observed to vary between natives and ethnic majorities and also between different minority groups themselves. Studies in the USA (Alba and Logan 1992; Krivo 1995; Painter et al. 2001; DeSilva and Elmelech 2012) and the UK (Nygaard 2011) have found income to have mostly *stronger* positive effects among ethnic minorities, while an Australian study (Chua and Miller 2009) found weaker effects among immigrants. Stronger income effects imply that becoming a homeowner is more dependent on higher income among immigrants than among natives, pointing towards that especially low-income immigrants face barriers against moving to homeownership or prefer to use their financial resources for other purposes. The Nordic study referenced later reaches, however, the opposite conclusion, that increasing income has a weaker effect for ethnic minorities. This might be due to that rented housing is more available in these countries.

Regarding the demographic determinants, the number of children in the household has been found to be more *weakly* connected to homeownership among ethnic minorities (Alba and Logan 1992; Chua and Miller 2009; Nygaard 2011; DeSilva and Elmelech 2012), suggesting either more constraints in the realisation of homeownership plans or lower preferences for homeownership. Results concerning marital status are more mixed, with weaker effects among immigrants or ethnic minorities in the US (Alba and Logan 1992; Krivo 1995; DeSilva and Elmelech 2012) and in the Nordic countries (Kauppinen et al. 2015) but stronger effects in the UK (Nygaard 2011) and Australia (Chua and Miller 2009).

Existing research has mostly relied on cross-sectional designs. The results may be inaccurate in these designs, since the current housing conditions tend to reflect 'the lagged effects of past choices' (Painter 2000, p. 198). Longitudinal panel data offer the best chances for assessing the effects of socio-demographic determinants, enabling the analysis of the effects of changes in life circumstances on the entry to homeownership (Raya and Garcia 2012). For example, family *events* (Feijten and Mulder 2002) or *stability* of employment (Magnusson Turner and Hedman 2014; Skifter Andersen 2011) may be important determinants in addition to the current situation. The need for longitudinal analysis has indeed been long acknowledged in the research on the housing conditions of immigrants (e.g. Myers 1999; Özüekren and Van Kempen 2002; Painter 2000), but longitudinal research has so far been scarce. Magnusson Turner and Hedman (2014) studied the housing career over time of immigrants in Sweden. The Nordic study referenced later is also based on longitudinal data.

It is obvious that the extent to which ethnic minorities move to homeownership in different countries is dependent on the economic situation of ethnic minorities in general and the conditions in the housing market, as will be discussed in Chapter 6. It is, however, also influenced by differences across minority groups concerning the conditions discussed in the beginning of this section; their cultural traditions and their links to their country or origin. Many studies hypothesise that there is a link between homeownership and social and cultural integration or assimilation (Alba and Logan 1992; Borjas 2002), as it indicates long-term economic progress and plays a key role in providing financial security (Sinning 2010), and it may also indicate immigrants' degree of commitment to the host country (Constant et al. 2009). It is assumed that the gap in homeownership between natives and immigrants depends on ethnic group and length of stay in the country. Studies in Canada and Australia (Chua and Miller 2009; Kim and Boyd 2009) have shown, however, that the gap only relates to recent immigrants.

The German study of Constant et al. (2009) was based on survey data on immigrants' knowledge of the German language, preferred media, social relations and migration plans, which were used as indicators of integration. These indicators were shown to affect significantly the probability of homeownership, showing assimilated and integrated immigrants to be much more likely to own their place of residence. It was demonstrated that those with a stronger commitment to and identification with the host country were more able to move into homeownership for a given set of socio-economic and demographic characteristics and housing market conditions.

It is difficult to interpret the results from studies based on revealed preferences because other factors than socio-economic and ethnic variables influence choice of housing tenure. Besides differences across countries and cities concerning housing markets, and immigrants' options in these markets, tenure choice is also influenced by the preferences among ethnic minorities for what kind of neighbourhood they want and the way tenures are distributed across neighbourhoods.

Effects of preferences for neighbourhood

As will be discussed in Chapter 3, some ethnic minorities prefer to settle in neighbourhoods where they can find an ethnic social network, called ethnic enclaves. Such neighbourhoods have often been created in parts of the cities with less attractive housing dominated by an easy-to-access tenure, which often is renting. This commonly has constituted the basis for an initial influx of immigrants (Søholt 2007; Søholt and Astrup 2009). Processes of segregation, as will be discussed in Chapter 5, lead to the native majority tending to move away from such neighbourhoods and avoiding moving to them (Clark 1992; Quillian 2002; Bråmå 2006; Bolt et al. 2008). If the housing demand from the majority ethnic group decreases, it is easier for ethnic minorities to get access to these neighbourhoods, and prices and rents tend to be lower. Some of the minority residents in such neighbourhoods could choose to stay put in rented housing despite high preferences for homeownership. A study analysing moves to homeownership in three Nordic capitals (Kauppinen et al. 2015) did, however, only show small effects of living in multiethnic neighbourhoods, while living in social housing strongly reduced the propensity to move into homeownership, probably because immigrants had obtained a satisfactory housing situation here.

The significance of neighbourhood characteristics and differentiation of neighbourhoods as determinants of entry to homeownership have been studied less. Flippen (2001) has shown that in the US, there is a negative association between ethnic spatial segregation and minority homeownership. In Europe ethnic segregation is less pronounced, which could mean that there is a weaker association between segregation and homeownership rates. However, European housing markets are more complex because of public involvement in the housing sector, especially the existence of larger social housing sectors, which tend to provide better housing options in renting for immigrants. In some countries immigrant-dense neighbourhoods are dominated by social housing, which is especially favourable for ethnic minorities with lower incomes. Combined with the *enclave effect*, this might furthermore reduce the propensity for ethnic minorities to move into homeownership. A Danish qualitative study (Skovgaard Nielsen 2017) thus showed that many immigrants found the economic conditions in social housing in Denmark so favourable that they did not consider buying a home. A comparative study of four Nordic countries (Skovgaard Nielsen et al. 2015) showed that this effect varied considerably across the countries depending on accessibility and conditions in social housing.

A study of entry into homeownership in three Nordic capitals

Content of the study

The purpose of the study was to compare the extent to which ethnic minorities move into homeownership with that of the native majority in three cities, and if homeownership is determined by other factors than found among natives, especially the role of income and income changes.

This study differs from earlier studies in two aspects. Firstly, it is based on longitudinal data on individuals, which makes it possible to follow the housing careers and examine the connection between moving to homeownership and changes in life situation. Secondly, it has data from three Nordic countries – Denmark, Finland and Sweden – that in many ways are very comparable, but which have a different amount and composition of immigrants and to some extent different housing markets and urban structures. This extends the research to other national contexts beyond the previously rather Anglo-American focused research. It also permits some assessment of the importance of the differing immigration histories and housing markets in the countries.

The study has been reported in detail in Kauppinen et al. (2015). We analyse entry to homeownership in the capital regions of the three countries: Copenhagen, Helsinki and Stockholm. Our starting point for each city region is to look at how big the differences are between natives and immigrants in the speed of progression to homeownership and to what extent the simplest explanation, i.e. different economic and demographic compositions, is enough to explain these differences.

We then proceed to our main research questions. The first of them is: are the effects of economic and demographic characteristics on the entry to homeownership different among natives and immigrants? The second main question is: do the ethnic composition of the present neighbourhood, and tenure type of present housing, influence the propensity to move into homeownership, even when the economic and demographic characteristics of households and individuals are controlled for? The differences in the immigration history between Finland and the other two countries can be expected to influence the conditions for homeownership. An important difference is that the shares of immigrants in the neighbourhoods are generally lower in Helsinki, and there are fewer neighbourhoods with high shares of immigrants.

The housing markets in the cities

The results of this study are by nature very dependent on housing policy and housing markets in the three countries. This will be described in more detail in Chapter 6. The housing tenure distributions for the total population and among immigrants in the three capital regions are illustrated in Table 2.1. The table also shows the under-representation in homeownership of immigrants, calculated as the relative difference in the homeownership rate between an immigrant group and the total population.

The proportion of the population living in homeownership is highest in Helsinki, followed by Copenhagen. However, if one includes cooperatives, Stockholm has the highest ownership rate. This holds also for immigrants, whose homeownership rate is the highest in Stockholm, and the under-representation of homeownership is the lowest of the three cities. The lowest homeownership rate among immigrants and the largest under-representation of homeownership is observed in Helsinki. A majority of immigrants in Helsinki are living in social housing.

Data and methods

Register-based administrative data is used. A separate longitudinal individual-level data set was constructed for each country. The data represent persons who have lived in the capital areas of the countries for at least one year between 1991 and 2000, being between 25 and 34 years of age. People are followed annually from the first year when they are observed in the metropolitan area between the years 1991 and 2000 as: 1. 25–34 years old, and 2. Not living in owner-occupation

Table 2.1 Housing tenure distributions (%) in the capital regions in 2008

	Copenhagen	*Helsinki*	*Stockholm*
Total population			
Owner-occupied	42	57	34
Cooperatives	16	3	29
Private renting	16	16	17
Social/public housing	25	22	17
Other	1	2	4
Total	100	100	100
All immigrants			
Owner-occupied	25	24	18
Cooperatives	13	2	25
Private renting	21	19	22
Social/public housing	39	52	30
Other	3	3	5
Total	100	100	100
Non-Western immigrants			
Owner-occupied	19	18	14
Cooperatives	8	0	20
Private renting	14	16	24
Social/public housing	58	62	41
Other	1	4	0
Total	100	100	100
Under-representation in homeownership, %			
All immigrants	−40	−58	−32
Non-Western immigrants	−55	−68	−46

Sources: Government statistical offices: Statistics Denmark, Statistics Finland, and Statistics Sweden

Note: Tenure distributions in the Helsinki region are by native language instead of country of birth as in Copenhagen and Stockholm. In the calculation of under-representation of homeownership, cooperatives are included as homeownership in Stockholm. For the definition of Non-Western immigrants, see Kauppinen et al. 2015.

on their own. More details on the selection of observations can be found in Kauppinen et al. (2015).

The dependent variable in the statistical analyses is entry to homeownership, which is observed when the housing tenure is owner-occupation for the first time during the course of time. We control for the number of years since the first year, which is either the year of turning 25 or the year of moving to the capital region. Immigrant status is measured by the country of birth. We have grouped countries into: Eastern Europe (the former Eastern Bloc countries, including Russia and the new EU members), Africa and West Asia (including Turkey), Other Asia, and (in Denmark and Sweden) Latin America. Immigrants from other countries are called Western immigrants. We measure several demographic characteristics: the age and sex of the person, having children in the family and living together with a partner. These variables were measured in a dynamic way, considering the recent changes. Socio-economic characteristics were measured with three variables in the analyses for the first main research question: the employment status in the current and the two previous years, the previous year's gross household income and the change of income between the previous and the current year. Household income was measured as the income decile within the 25–44-year-old population in the capital region. The change of income was measured by dividing the current (inflation-corrected) income by the previous year's income.

In the analyses of the importance of present housing situation, we used a variable measuring the percentage of Non-Western immigrants in the previous year neighbourhood's population. We also used the percentage of owner-occupied dwellings in the housing stock of the neighbourhood. Another additional variable was if the previous year's housing tenure was social housing.

The methods used are described in more detail in Kauppinen et al. (2015).

Comparison of the speed of progression to homeownership

We first created reversed survival graphs in order to show the differences between country-of-birth groups in the speed of entry to homeownership. These figures were based on life tables constructed for synthetic cohorts, i.e. combining processes starting in different calendar years. Figure 2.1 shows for each capital region, how quickly the country-of-birth groups enter homeownership. The number of years since entry measures time since the start, which is either the age of 25 or, for those who arrived at a later age to the city region, the year of the arrival.

These figures illustrate that the difference between natives and the slowest progressing immigrant group – Africa and West Asia – is roughly similar in the three regions. The relative speed of entering homeownership among the other ethnic groups is also quite similar across the cities with a few exceptions. In Copenhagen and Stockholm immigrants from East Europe and Other Asia follow the same pattern, while East European immigrants in Helsinki have a slower progression than immigrants from Other Asia. An explanation could be that East European immigrants in Helsinki mostly come from countries outside the EU.

Comparison of economic and demographic determinants of entering homeownership

We started the analysis by running logistic models that control for the basic demographic and economic variables like income, employment, family situation, ethnic group, etc. Results in detail are reported in Kauppinen et al. (2015). Figure 2.2 shows the marginal effects measured in models for the three cities containing both natives and immigrants.

The results for these three models show that differences between natives and immigrants cannot be explained by differences in the socio-demographic composition across natives and immigrants. A notable result is also that in spite of the major differences in the housing markets across the cities, the differences between specific immigrant groups and natives are remarkably similar. In all three cities nearby the same difference from the natives is found for the three Non-Western

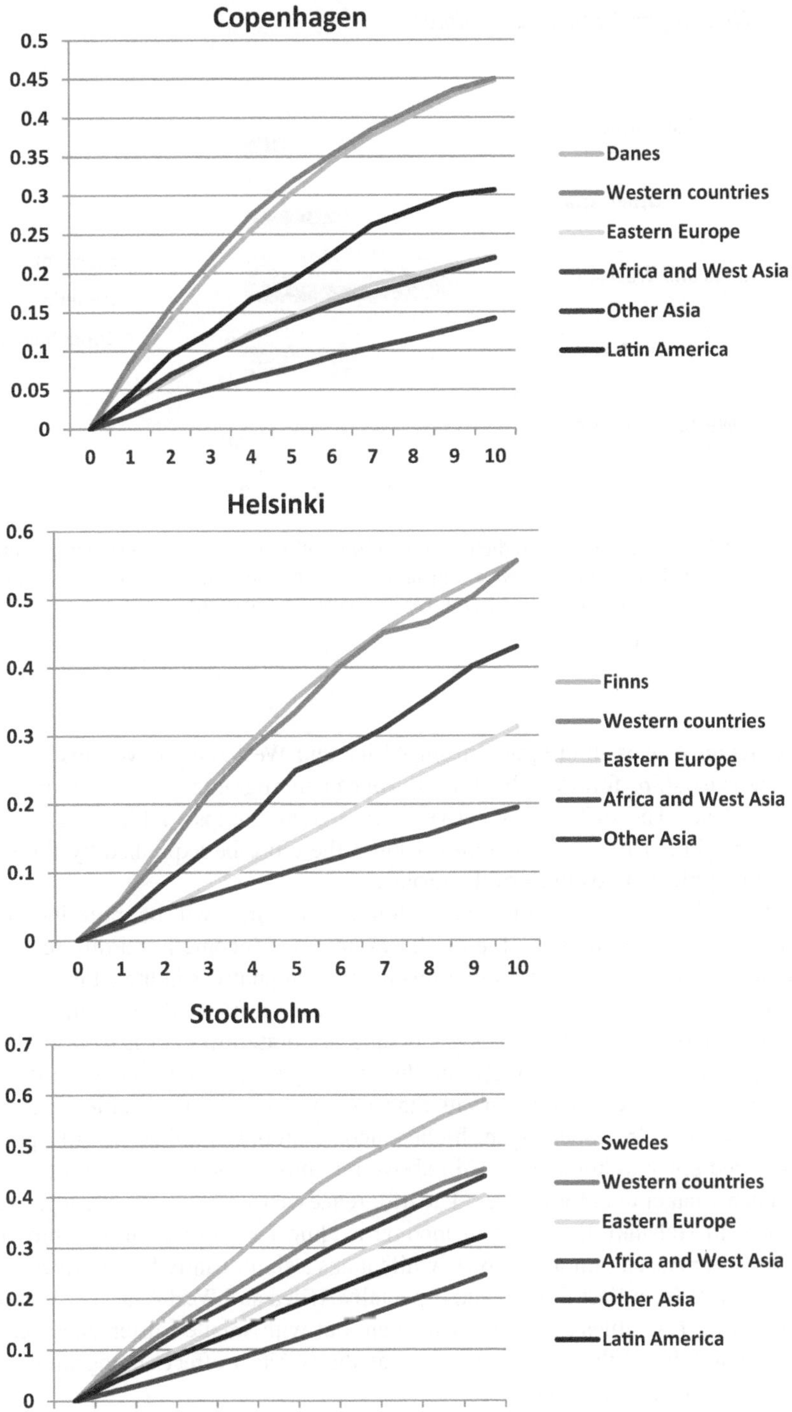

Figure 2.1 Survival curves for ethnic minorities for what year after arrival to leave rental housing for homeownership compared with natives in Copenhagen, Helsinki and Stockholm

Source: Kauppinen et al. 2015

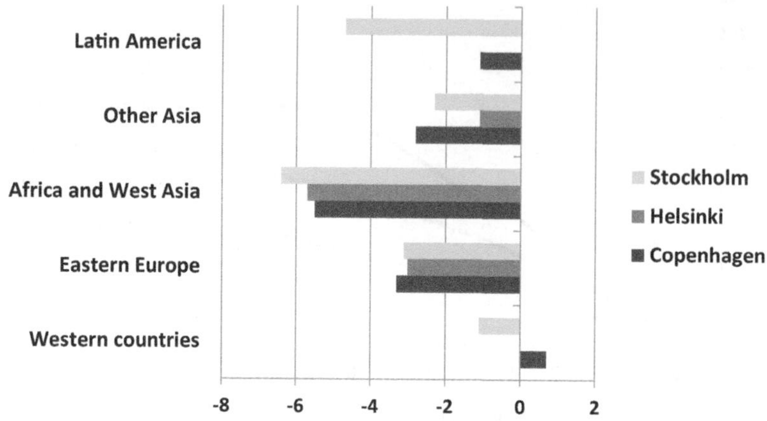

Figure 2.2 Differences in predicted probability of entering homeownership between different *ethnic groups* and natives in Copenhagen, Helsinki and Stockholm (marginal effects (%-points) from logistic regression)

Source: Kauppinen et al. 2015

immigrant groups. Immigrants from Africa and West Asia move most rarely to homeownership, followed by East European immigrants and immigrants from Other Asia. The differences between immigrant groups in Figure 2.1, which also are similar between the cities, cannot therefore be explained by the socio-demographic compositions of the groups.

It is to be expected that people with higher incomes will be more inclined to move into homeownership. Figure 2.3 examines if income has a different effect on homeownership among immigrants than among natives in the cities.

In all three cities household income has an important effect on moving into homeownership, but the effects of income are quite different across immigrant groups and cities. The models for the ethnic groups show that the effects of income are weaker for immigrants than for natives in all three cities, but not in the same way. In Copenhagen, higher income among immigrants only predicts increased entry to homeownership above the third income decile, and the effect is much weaker than for natives. The differences across immigrant groups are not strong. In Helsinki, the effect of income is almost as strong among immigrants from Other Asia as among natives, while it has a significantly lower effect among immigrants from East Europe and especially Africa and West Asia. The pattern in Stockholm resembles that in Copenhagen, but with smaller differences between the groups. In all three cities income has the weakest effect among immigrants from Africa and West Asia.

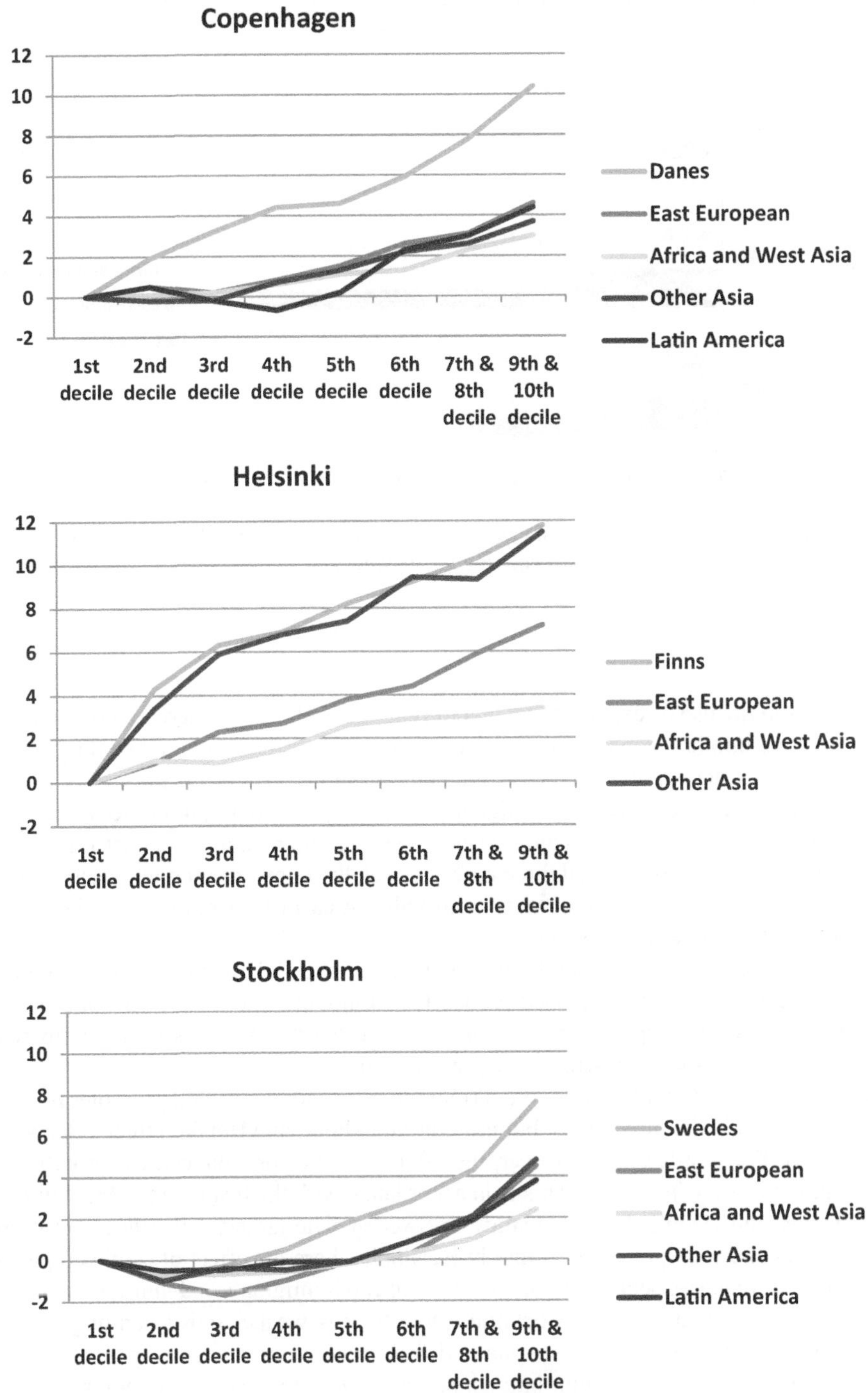

Figure 2.3 Differences in predicted probability (marginal effects %-points) of entering homeownership between different *income groups* (deciles) and the lowest incomes (decile 1) for different ethnic groups in the three cities

Source: Kauppinen et al. 2015

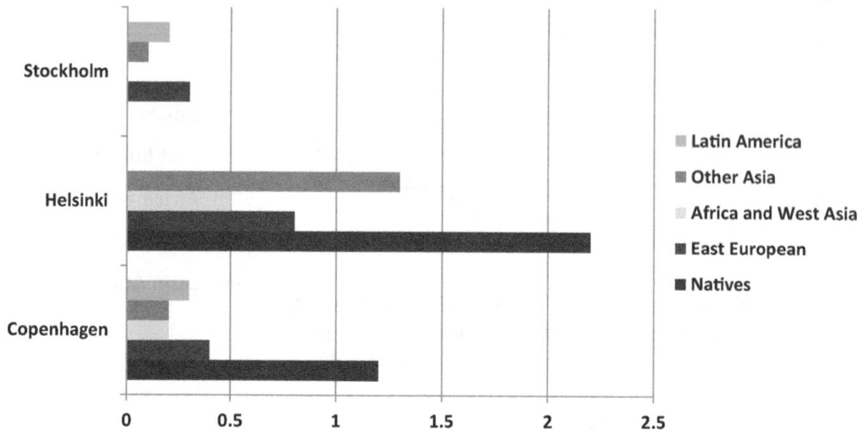

Figure 2.4 Effects of changes in income since last year on predicted probability (marginal effects %-points) of entering homeownership

Source: Kauppinen et al. 2015

It should also be expected that income changes will trigger increased propensity to move into homeownership. Figure 2.4 shows the effects on predicted mobility to homeownership of the change in income from last year to the present year.

Increased income results in higher probability for moving to homeownership, mostly among natives. Income changes have a much smaller impact on immigrants than on natives in all three cities, especially among immigrants from Africa and West Asia. Only immigrants from Other Asia in Helsinki seem to be more affected by increasing income.

As discussed earlier, stable employment is essential for purchasing a home. Figure 2.5 shows the increased probability of moving into homeownership when having been in employment now and at least the two previous years compared with being not in employment the present year.

As expected, stable employment (more than 2 years) in general promotes homeownership in the three cities, but more in Stockholm and Helsinki than in Copenhagen. For immigrants, however, the effect is not as pronounced as for natives, but is still very important. This is in accordance with the results from Magnusson Turner and Hedman's (2014) study in Sweden. The largest differences between natives and immigrants are found in Helsinki, where the effect of stable employment is weaker especially among East European immigrants and immigrants from Africa and West Asia. In Stockholm the effect is weaker mostly among immigrants from Africa and West Asia and from Latin America. In Copenhagen the effects are small among immigrants, but quite close to the effect among natives.

Gaining employment also affects entry to homeownership (Figure 2.6).

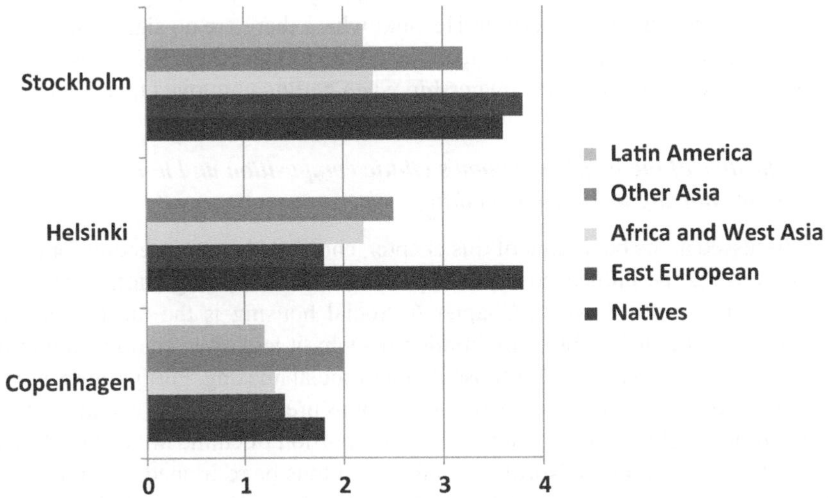

Figure 2.5 The effect on predicted probability (marginal effects %-points) of moving to homeownership of having *been in employment* at least 2 years compared to not being in employment for the last 2 years

Source: Kauppinen et al. 2015

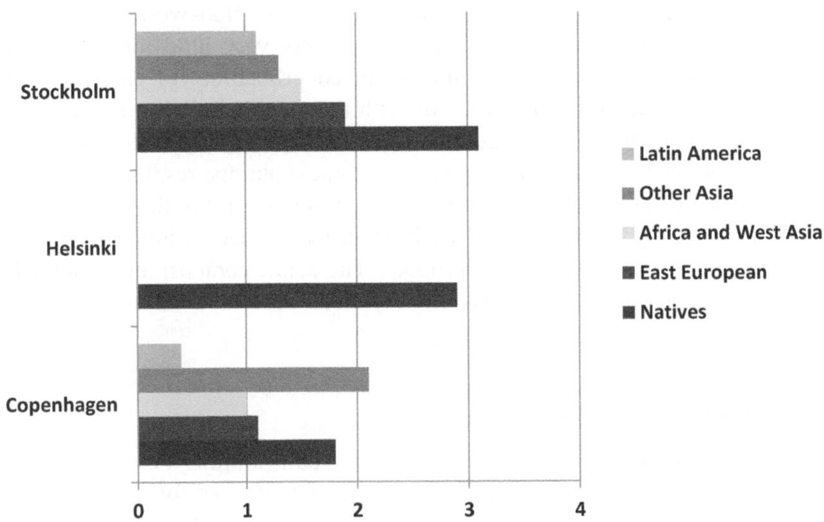

Figure 2.6 The effect on predicted probability (marginal effects %-points) of moving to homeownership of *getting a job in present year* compared to not being in employment for the last 2 years

Source: Kauppinen et al. 2015

The effect of obtaining employment is mostly smaller among immigrants in all three cities, but most clearly in Helsinki where there are no significant effects for immigrants. In Copenhagen one of the ethnic groups, Other Asia, has a higher propensity to move into homeownership when getting employed than Danes.

Significance of the neighbourhood's ethnic composition and housing tenure for moving to homeownership

As discussed at the beginning of this chapter, choice of homeownership for ethnic minorities can be influenced by their position on the housing market. This will be treated in more detail in Chapter 6. Social housing is the most option for getting decent housing for many immigrants why it less often could be attractive to move to homeownership for those living in social housing. Furthermore, as will be discussed in Chapter 3, some minority groups prefer living close to their ethnic network in neighbourhoods with a higher proportion of ethnic minorities. Moves to homeownership in the Nordic capitals might thus be restrained for immigrants living in multi-ethnic neighbourhoods. In the Nordic countries, spatial concentrations of ethnic minorities are mostly found in social housing.

In all three cities, those living in social housing are less likely to enter homeownership. In Stockholm and especially in Copenhagen this effect is, however, stronger among natives, but in Helsinki it is similar or stronger among immigrants, although not significant in all groups. Among immigrants from Africa and West Asia in Helsinki, the effect is also strong in comparison with the other explanatory variables.

The statistical results indicate that the percentage of Non-Western immigrants in the former neighbourhood's population matters very little for moving into homeownership. In Copenhagen, the predicted probability of entering homeownership is lowest in neighbourhoods with the most Non-Western immigrants, but the difference from other types of neighbourhoods is very small. In Helsinki, the effects are not statistically significant. In Stockholm the results are otherwise similar to those in Copenhagen, but for most immigrant groups the entry to homeownership actually seems to increase a little, instead of decreasing, when living in the most immigrant-dense neighbourhoods. The ethnic composition of the neighbourhood population seems to have slightly more importance among immigrants from Africa and West Asia.

Findings from the Nordic study

The referenced Nordic study discovered that ethnic minorities more seldom than natives move to homeownership even when controlled for socio-economic and demographic background variables. The differences in homeownership rate between different ethnic groups were quite alike across the cities in spite of some differences in the national contexts regarding level and composition of immigration, housing markets and economic conditions. This indicates that cultural factors and the special position in society for immigrants have an essential importance for homeownership across countries in spite of major differences in local conditions.

The study revealed a difference from earlier findings from the USA and the UK: the effect of increasing income on homeownership in the Nordic capitals is *weaker* – or at least not stronger – among immigrants than among natives. Moreover, experiences with stable employment were important. This supports the assumption of higher uncertainty regarding future economic circumstances, given the difficulties in employment. It may also reflect higher uncertainty of expectations of future length of stay in the country, especially among immigrants with refugee backgrounds. The Nordic study also discovered that ethnic minorities more seldom than natives move to homeownership in connection with family changes, also when controlled for income.

Some studies have concluded that ethnic segregation and the way multi-ethnic neighbourhoods have occurred in the countries, and the tenures they contain, affects tenure choice of ethnic minorities. A significant conclusion from the Nordic study is, however, that increased ethnic segregation does not seem to reduce homeownership in the three examined Nordic cities. This differs from the findings in the US and may be due to weaker segregation and less institutionalised separation of ethnic groups in the Nordic context.

Note

1 National Institute for Health and Welfare, Helsinki, Finland. Lina Hedman, Institute for Housing and Urban Research, University of Uppsala, Sweden, has contributed with data from Stockholm.

References

Alba R. D. and Logan J. R. (1992) Assimilation and Stratification in the Homeownership Patterns of Racial and Ethnic Groups, *International Migration Review*, 26(4), pp. 1314–1341.

Bolt G., Van Kempen R. and Van Ham M. (2008) Minority Ethnic Groups in the Dutch Housing Market: Spatial Segregation, Relocation Dynamics and Housing Policy, *Urban Studies*, 45(7), pp. 1359–1384.

Borjas G. J. (2002) Homeownership in the Immigrant Population, *Journal of Urban Economics*, 52(3), pp. 448–476.

Bowes A., Dar N. and Sim D. (1997) Tenure Preference and Housing Strategy: An Exploration of Pakistani Experiences, *Housing Studies*, 12(1), pp. 63–84.

Bråmå Å. (2006) 'White Flight'? The Production and Reproduction of Immigrant Concentration Areas in Swedish Cities 1990–2000, *Urban Studies*, 43(7), pp. 1127–1146.

Bråmå Å. And Andersson R. (2010) Who Leaves Rental Housing? Examining Possible Explanations for Housing Segmentation in Uppsala, Sweden, *Journal of Housing and the Built Environment*, 25(3), pp. 331–352.

Chua J. and Miller P. W. (2009) The Impact of Immigrant Status on Home Ownership in Australia, *International Migration*, 47(2), pp. 155–192.

Clark W. A. V. (1992) Residential Preferences and Residential Choices in a Multi-ethnic Context, *Demography*, 29(3), pp. 451–466.

Clark W. A. V., Deurloo M. C. and Dieleman F. M. (1997) Entry to Homeownership in Germany: Some Comparisons with the United States, *Urban Studies*, 34(1), pp. 7–19.

Clark W. A. V. and Onaka J. L. (1983) Life Cycle and Housing Adjustment as Explanations of Residential Mobility, *Urban Studies*, 20(1), pp. 47–57.

Constant A. F., Roberts R. and Zimmermann K. F. (2009) Ethnic Identity and Immigrant Homeownership, *Urban Studies*, 49(6), pp. 1879–1898.

Coolen H. and Hoekstra J. (2001) Values as Determinants of Preferences for Housing Attributes, *Journal of Housing and the Built Environment*, 16, pp. 285–306.

Davidov E. and Weick S. (2011) Transition to Homeownership Among Immigrant Groups and Natives in West Germany, 1984–2008, *Journal of Immigrant & Refugee Studies*, 9(4), pp. 393–415.

Desilva S. and Elmelech Y. (2012) Housing Inequality in the United States: Explaining the White-Minority Disparities in Homeownership, *Housing Studies*, 27(1), pp. 1–26.

Dieleman F. M., Clark W. A. W. and Deurloo M. C. (1998) Tenure Choice: Cross-sectional and Longitudinal Analyses, *Netherlands Journal of Housing and the Built Environment*, 9(3), pp. 229–246.

Feijten P. and Mulder C. H. (2002) The Timing of Household Events and Housing Events in the Netherlands: A Longitudinal Perspective, *Housing Studies*, 17(5), pp. 773–792.

Flippen C. A. (2001) The Spatial Dynamics of Stratification: Metropolitan Context, Population Redistribution, and Black and Hispanic Homeownership, *Demography*, 47(4), pp. 845–868.

Floor H. and Van Kempen R. (1997) Analysing Housing Preferences with Decision Plan Nets, *Scandinavian Housing and Planning Research*, 14, pp. 27–42.

Haan M. (2007) The Homeownership Hierarchies of Canada and the United States: The Housing Patterns of White and Non-White Immigrants of the Past Thirty Years, *International Migration Review*, 41(2), pp. 433–465.

Howell F. M. and Frese W. (1983) Size of Place, Residential Preferences and the Life Cycle: How People Come to Like Where They Live, *American Sociological Review*, 48(4), pp. 569–580.

Kauppinen T., Skifter Andersen H. and Hedman L. (2015) Determinants of Immigrants' Entry to Homeownership in Three Nordic Capital City Regions, *Geografiska Annaler: Series B, Human Geography*, 97(4), pp. 1–20.

Kim A. H. and Boyd M. (2009) Housing Tenure and Condos: Ownership by Immigrant Generations and the Timing of Arrival, *Canadian Journal of Urban Research*, 18(1), pp. 47–73.

Krivo L. J. (1995) Immigrant Characteristics and Hispanic-Anglo Housing Inequality, *Demography*, 32(4), pp. 599–615.

Magnusson Turner L. and Hedman L. (2014) Linking Integration and Housing Career: A Longitudinal Analysis of Immigrant Groups in Sweden, *Housing Studies*, 29(2), pp. 270–290.

Mensah J. and Williams C. J. (2014) Cultural Dimensions of African Immigrant Housing in Toronto: A Qualitative Insight, *Housing Studies*, 29(3), pp. 438–455.

Mulder C. H. (2007) The Family Context and Residential Choice: A Challenge for New Research, *Population, Space and Place*, 13(4), pp. 265–278.

Myers D. (1999) Cohort Longitudinal Estimation of Housing Careers, *Housing Studies*, 14(4), pp. 473–490.

Nygaard C. (2011) International Migration, Housing Demand and Access to Homeownership in the UK, *Urban Studies*, 48(11), pp. 2211–2229.

Owusu T. Y. (1998) To Buy or Not to Buy: Determinants of Home Ownership among Ghanaian Immigrants in Toronto, *Canadian Geographer*, 42(1). Pp. 40–52.

Özüekren A. S. and Van Kempen R. (2002) Housing Careers of Minority Ethnic Groups: Experiences, Explanations and Prospects, *Housing Studies*, 17(3), pp. 365–379.

Painter G. (2000) Tenure Choice with Sample Selection: Differences among Alternative Samples, *Journal of Housing Economics*, 9(3), pp. 197–213.

Painter G., Gabriel S. and Myers D. (2001) Race, Immigrant Status, and Housing Tenure Choice, *Journal of Urban Economics*, 49(1), pp. 150–167.

Painter G. and Yu Z. (2010) Immigrants and Housing Markets in Mid-Size Metropolitan Areas, *International Migration Review*, 44(2), pp. 442–476.

Phillips D. (2009) Creating Home-spaces: Young British Muslim Women's Identity and Conceptualisations of Home. In Hopkins P. and Gale R. *Muslims in Britain: Race, Place and Identities*, pp. 3–36. Edinburgh: Edinburgh University Press.

Portes A. and Sensenbrenner J. (1993) Embeddedness and Immigration: Notes on the Social Determinants of Economic Action, *American Journal of Sociology*, 98(6), pp. 1320–1350.

Quillian L. (2002) Why is Black – White Residential Segregation So Persistent? Evidence on Three Theories from Migration Data, *Social Science Research*, 31(2), pp. 197–202.

Raya J. and Garcia J. (2012) Which Are the Real Determinants of Tenure? A Comparative Analysis of Different Models of the Tenure Choice of a House, *Urban Studies*, 49(16), pp. 3645–3662.

Rossi P. H. (1955) *When Families Move*. Glencoe, IL: The Free Press.

Sinning M. (2010) Homeownership and Economic Performance of Immigrants in Germany, *Urban Studies*, 47(2), pp. 387–409.

Skifter Andersen H. (2011) Motives for Tenure Choice During the Life Cycle. The Importance of Non-economic Factors and Other Housing Preferences, *Housing, Theory and Society*, 28(2), pp. 183–207.

Skifter Andersen H. (2006) *Bo sammen eller spredt? Etniske minoriteters boligønsker og motiver for bosætning* (Living Together or Separated. Preferences for Settlement among Ethnic Minorities). Horsholm: Danish Building Research Institute.

Skifter Andersen H. and Bonke J. (1980) *Boligsektorens fordelingsmæssige virkninger*. Copenhagen: Lavindkomstkommissionen.

Skovgaard Nielsen R. (2017) The Potentials of a Strong Social Housing Sector: The Case of Turks and Somalis in the Copenhagen Housing Market, *Housing, Theory and Society*, 34(4), pp. 458–476.

Skovgaard Nielsen R. (2016) Straight-line Assimilation in Leaving Home? A Comparison of Turks, Somalis and Danes, *Housing Studies*, 31(6), pp. 631–650.

Skovgaard Nielsen R. (2015) Leaving Home in Denmark: A Comparison of Differences between Turks, Somalis and Danes, *Journal Housing and the Built Environment*, 30, pp. 613–630.

Skovgård Nielsen R., Holmquist E., Dhalmann H. and Søholt S. (2015) Comparing Somalis' Perceived Possibilities in the Nordic Capital Housing Markets: Local Context and Cultural Background, *Housing Studies*, 30(3), pp. 433–452.

Søholt S. (2014) Pathways to Integration: Cross-cultural Adaptations to the Housing Market in Oslo, *Journal of Ethnic and Migration Studies*, 40(10), pp. 1637–1656.

Søholt S. (2007) *Gjennom nåløyet – en sammenligning av tilpasninger til boligmarkedet blant hushold av pakistansk, tamilsk og somalisk bakgrunn, Oslo 1970–2003*. Thesis. Institute for Political Science, Faculty of Social Sciences, University of Oslo, Oslo.

Søholt S. and Astrup K. (2009) *Etniske minoriteter og forskjellsbehandling i leiemarkedet*. NIBR-report: 2. Norwegian Institute for Urban and Regional Research, Oslo.

Timmermanns H., Molin E. and van Noortvijk L. (1994) Housing Choice Processes: Stated versus Revealed Modelling Approaches, *Journal of Housing and the Built Environment*, 9(3), pp. 215–227.

Toussaint-Comeau M. (2012) Ethnic Social Networks and Self-employment of Immigrant Men in the US, *Eastern Economic Journal*, 38, pp. 74–98.

Zorlu A. and Mulder C. (2011) Ethnic Differences in Leaving Home: Timing and Pathways, *Demography*, 48, pp. 49–72.

3 Preferences for location among ethnic minorities

Chapter 2 describes how ethnic minorities' housing preferences sometimes differ from those of the native majority. This has an impact on ethnic segregation if different kinds of housing are spatially separated in the cities, as described in Chapters 6 and 7. The location of ethnic minorities might, however, also be a result of having preferences for location which differ from those of natives and other ethnic groups.

In general, preferences for location concern dimensions such as: 1. The physical environment, 2. The social and cultural environment, 3. Access to local public and private service facilities, and 4. Location and transport in a broader context (Skifter Andersen 2011). It has been shown that preferences vary greatly between people, and that this variation mostly depends on differences in family situation and course of life, income and other resources, and culture and lifestyle (Ærø 2006).

Research on locational preferences among ethnic minorities has not been much concerned with differences across ethnic groups concerning their propensity towards different kinds of physical environments or proximity to transport. Most of the research has focused on the social and cultural environment and to some extent service facilities. The absolute main focus has been on to what extent ethnic minorities prefer to live close to neighbours from their own ethnic group, subsequently if they want to avoid living in neighbourhoods dominated by the native majority or by other groups. In the literature this has often been named propensity to *self-segregation* (Phillips 2006; Van Der Laan Bouma-Doff 2007; Dhalman 2013).

The literature on this is vast and conclusions have varied dependent on country and the groups studied. Most studies have been based on quantitative data on where people actually live, so-called *revealed preferences* (see the discussion in Chapter 2). Only a few have used surveys or qualitative methods where people are asked directly about their preferences.

In principle, three different kinds of preferences can be formulated: 1. To live close to family networks and friends, 2. To locate in neighbourhoods with many countrymen, or ethnic minorities with a similar culture, to become part of a local ethnic social and economic network, or 3. To avoid living alone among the native majority, or other groups, to escape discrimination or racism.

The first kind is not very different from preferences among the native majority, but it can contribute to a concentration of ethnic minorities in certain neighbourhoods if it is more important for immigrants than for natives, and if their options on the housing market are more restricted to certain neighbourhoods. The second is the most often mentioned motive in the literature. A basic standpoint is that moving to neighbourhoods with an ethnic network, sometimes called *ethnic enclaves*, is an important motive among especially newly arrived immigrants as a strategy for survival and integration in their new country, because they can get support for finding housing and jobs. Often there are also local shops that purchase consumer goods from the homeland, which can ease the access and reduce the costs of using special ethnic goods and services. The third preference could be motivated by the search for security and safety in a well-known social and cultural environment, and to escape what is experienced as persecution and discrimination from the native majority.

There is considerable disagreement about the importance of these preferences for segregation. As most studies have used quantitative data on the actual location of ethnic minorities, it is difficult to decide if an ethnic spatial concentration is due to preferences or to lack of options on the housing market. A few studies based on surveys and some qualitative research exist, which can qualify the discussion. Results from a survey among moving ethnic minorities in Denmark, referenced in the last section of this chapter, indicate that the first of the three mentioned kinds of preferences is the most important.

We will start the chapter with a review of the literature. In the last part of the chapter the results from a study on ethnic minorities' preferences for location in Denmark will be described, and how such preferences are influenced by factors like ethnic background, economic resources and degree of social integration in the Danish society will be illuminated.

Knowledge on special preferences for location among ethnic minorities

The American debate on ethnic enclaves *and* ghettos

In the US it has been observed from the start of immigration in the 19th century that new immigrants to the country tended to settle in neighbourhoods with many countrymen. According to the American 'ecological' theoretical tradition of the Chicago School (Park 1925), this was a natural process where immigrants could get support from other immigrants coming from the same country and manage by speaking their mother tongue. As stated by Massey (1985): 'Some degree of geographic concentration is an inevitable by-product of immigration, which is guided by social networks and leads to settlement patterns determined partly by the need of new immigrants unfamiliar with American society and frequently lacking proficiency in English for assistance from kin and co-ethnics'. This process has been supported by so-called *chain migration processes* (Johnston et al. 2002), where immigrants who have arrived earlier have drawn a number of newer immigrants to the host country through kinship and other social networks.

In the literature two different terms, *ghettos* and *ethnic enclaves*, have been used to refer to neighbourhoods with a high concentration of a certain ethnic group or with ethnic minorities in general. Especially in European research, the two concepts sometimes have been confused and have been used to refer to neighbourhoods with a mix of different ethnic groups.

The concept *ghetto* has mainly been used to characterise neighbourhoods where one or several ethnic minorities have been concentrated as a consequence of practices of discrimination and exclusion, fostering both geographic and social isolation (Massey and Denton 1993; Marcuse 1997; Qadeer and Kumar 2006). Residential segregation is a necessary but not sufficient condition for a ghetto. In principle ghettos are not created because of preferences for living together among ethnic minorities but by strong limitations on their options for where to settle. Poverty, deprivation and discrimination are the defining conditions of a ghetto. They are places with low status and poor and deteriorated housing where no one wants to live except for the mostly deprived among ethnic minorities. The term ghetto is most often used on poor urban neighbourhoods in the US with a concentration of African-Americans. As stated by Wacquant (1999), Peach (1998) and Malheiros (2002), neighbourhoods in Europe cannot be characterised as ghettos because they mostly consist of many different kinds of minorities and as some of the residents have chosen to live there voluntarily.

The term *ethnic enclave* was originally used to characterise neighbourhoods that not only provided housing for a special ethnic group, but also employment possibilities provided by local ethnic businesses. This topic was at first described in an article by Wilson and Portes (1980) on newly arrived Cuban refugees and their labour market experiences in Miami from 1973 to 1976. Wilson and Portes found that a sizeable proportion of the newcomers went to work for co-ethnics having businesses in the same neighbourhood. They also discovered that those who worked for these immigrant firms were doing better than refugees employed in white-owned, secondary sector firms outside the neighbourhood. This gave birth to an extensive literature on the importance of ethnic enclaves for the employment of immigrants in the US and Canada (see the overview in Waldinger 1993, who also provides a critique of the thesis of better employment possibilities in the enclave). An ideal model for ethnic enclaves has been the so-called Chinatowns inhabited by Chinese immigrants. Research on those has mostly been made in North America (Zhou 1992; Kinkead 1992; Hatziprokopiou and Montagna 2012) but also in Britain (Luk 2008; Barabantseva 2016).

Ethnic enclaves have thus in this understanding been seen as very attractive for new immigrants. As stated by Portes and Manning (1986):

> *Once an enclave economy has fully developed, it is possible for a newcomer to live his life entirely within the confines of the community. Work, education, and access to health care, recreation, and a variety of other services can be found without leaving the bounds of the ethnic economy. This institutional completeness is what enables new immigrants to move ahead economically, despite very limited knowledge of the host culture and language.*

The use of the concept ethnic enclaves has later in the American literature been extended also to encompass kinds of neighbourhoods without major employment possibilities but with a high proportion of residents coming from the same ethnic group (Marcuse 1997; Abrahamson 1995 and 2004; Fong and Chan 2010). Qadeer and Kumar (2006) state:

> *Ethnic concentrations may come in many forms. A cluster of households of one ethnicity in a building or street is a small and unobtrusive agglomeration of ethnics (in a neighbourhood). When a particular ethnic group forms a large proportion of a neighbourhood's population, it becomes a geographic concentration. Mere living side by side without any community bonds and shared sentiments does not make an ethnic neighbourhood. It is the emergence of formal and informal community institutions and symbols that converts a concentration into an ethnic neighbourhood and eventually an enclave. Formally, a residential enclave is an area where a particular ethnic group numerically dominates, and has spawned corresponding religious, cultural, commercial and linguistic services and institutions. An enclave is a culturally and economically distinct area.*

According to Abrahamson (1995), they can be defined as places that tend to be relatively self-contained institutionally; that is, there are usually institutions attached to the enclave that support people's distinctive ways of life, such as schools that teach in their native language or homes for the aged that accommodate cultural dietary preferences. If the community is large enough, it can also have a same-language newspaper, radio or television station located within the enclave. Because so much of the lives of residents revolves around the enclave, their identities are frequently associated, such that members of an ethnic, religious or racial group sharing common traditions support specialised shopping venues, such as ethnic groceries or religious goods stores. For some immigrants it can also be important that enclaves sustain cultural connections with the home country through familiar language, social networks and ethnic organisations (Wright et al. 2005). This motive is connected to the concept of *diaspora*, which denotes the extent to which immigrants feel a split between their affiliation to the host country and their country of birth.

Ethnic enclaves can theoretically have two different models, according to Logan et al. (2002). One has sometimes been called the *immigrant enclave model*, in which ethnic neighbourhoods in central cities serve relatively impoverished new arrivals. The immigrant enclave model is in accordance with the theories on *spatial assimilation* proposed by the Chicago School (Massey and Denton 1985; Massey 1985; South et al. 2005). It is supposed that newly arrived immigrants will tend to move into neighbourhoods with many countrymen as part of a strategy for survival and integration in the new country. The immigrant enclave neighbourhoods are seen as temporary starting points for an immigrant's residential career. In accordance with the spatial assimilation theory, they are expected to leave the neighbourhoods when they have been more integrated, economically and socially, in their new country (see Chapter 4).

The other model, described by Logan et al. (2002), is named the *ethnic community model*, where segregated settlement can result from group preferences even when spatial assimilation is otherwise feasible. In some of the American literature enclaves – in contrast to ghettos – are not seen as areas of high concentration of poverty or deprivation even if an ethnic neighbourhood is easy to identify and stereotype and in extreme circumstances may become the target of prejudice and violence. Yet in an open society, the advantages of enclaves might outweigh the disadvantages. Such American enclaves are largely the product of market forces and personal choices.

Evidence from Canada has shown that some enclaves here have shifted from location in older neighbourhoods in central cities to suburbs (Qadeer and Kumar 2006; Fong and Chan 2010). Qadeer and Kumar (2006) claim that ethnic enclaves in Canada have a vibrant local commercial and services infrastructure. According to Alba and Nee (1997), new immigrant groups, mostly Asians and Hispanics, often from their arrival, or within 10 years, settled in suburbs in the US, where they to some extent formed kinds of ethnic enclaves. The American enclaves cannot, according to these sources, altogether be characterised as places of poverty and despair, and their social benefits outweigh the disadvantages of the predominance of one ethnic group. They find that the choice of ethnic households to live in areas of high concentration of their own group has two implications. First, living in such neighbourhoods can be viewed as beneficial on balance. There are distinct advantages of the critical mass of ethnics in a neighbourhood. Elderly and homebound women find companionship among those who speak their language and have many common interests and values. It facilitates the socialisation of children in their heritage culture. Politically and socially, minority communities feel strength in numbers. They can form voters blocks that politicians are compelled to pay heed to. Ethnic stores, services and places of worship become viable and emerge to enrich an area. Places of worship can be established close by. Organising religious and cultural activities is easy. Disadvantages may be the possibilities of children not being fully attuned to the mainstream values and being delayed in learning English.

The study conducted by Logan et al. (2002) of the residential patterns of the largest immigrant groups in New York and Los Angeles showed that most ethnic neighbourhoods here could be interpreted as immigrant enclaves. In some cases, however, living in ethnic neighbourhoods was unrelated to economic constraints, indicating a positive preference for such areas. Suburban residence did not necessarily imply living outside of ethnic neighbourhoods. For several groups the suburban enclave provided an alternative to spatial assimilation. It was an ethnic community in a relatively high-status setting.

The assumption that new immigrants and ethnic minorities in general have to live in the same neighbourhood to establish strong ethnic networks has been challenged in the literature on what has been called *heterolocalism* (Zelinsky and Lee 1998). Zelinsky and Lee based their theoretical considerations on a number of studies on the spatial behaviour of ethnic minorities in the US. They argue that advances in transport and communication technology have weakened the link between spatial concentration and social network formation resulting in the growth of ethnic communities without propinquity. Heterolocalism dismisses the relationship between spatial propinquity and ethnic communities, holding that such communities are nurtured by other means than mere physical proximity. Heterolocally inclined individuals may belong to ethnic communities without having a preference for spatial overlap between residence, social relations, workplace and sites of social activities. Qadeer and Kumar (2006) also argue that persons of one ethnicity may be a majority in a neighbourhood, but their work places, educational and health services, professional associations and social networks are mostly spread all across a city. It is therefore not appropriate to assume that ethnic composition of one's residential area defines the scope of one's social life. Educational institutions, political/economic organisations, professional groups and voluntary associations are the critical sites of social cohesion. Schools and universities, newspapers and mass media acculturate citizens in common values and national sentiments. All in all, it may be that enclaves symbolically tie together residents more than a typical neighbourhood; but they neither inhibit their contacts with the larger community, nor do they provide full range of facilities and services necessary for modern living.

Bushi (2014) used the theory of heterolocalism in a study of Albanian immigrants to the US, which showed that also recent immigrants showed behaviour that is comparable to the pattern described by Zelinsky and Lee. It can be argued, however, that Albanians in the US are such a small group that they do not have the capacity to form enclaves. Moreover, they are a group that does not deviate much from what is called the white majority in the US.

In the US there has been an intensive debate about the extent to which a preference among ethnic minorities for living in neighbourhoods dominated by their own ethnic group is a result of prejudices against other groups or fear of racial harassment from them. Most of the literature has focused on why the white majority avoids neighbourhoods with African-Americans (see Chapter 5), but some studies based on surveys have also examined the preferences of minorities (Clark 1992; Farley et al. 1997; Bobo and Zubrinski 1996).

The studies of Clark (1992) and Farley et al. (1997) showed that most African-Americans preferred mixed neighbourhoods, which did not only contain their own group but also residents belonging to the white majority. They did not want to live either in all-white neighbourhoods or neighbourhoods with their own group only. Different explanations have been put forward for these preferences. In the opinion of Clark (1992), they are a reflection of expressed desires to be with groups of ethnic neighbours of a similar background as one's self. He proposed that preferences are mostly driven by positive feelings about one's own group, not negative feelings about out-groups.

Farley et al. (1997), however, argue that racial prejudice and conflict underlie preferences of whites, and that blacks' preferences are largely a function of their concerns about the hostility they might face in a white neighbourhood. The situation of African-Americans in North America is a special case compared to other immigrant groups. Bobo and Zubrinski (1996) conducted a study that encompassed other ethnic minorities than African-Americans. They found little support for the importance of in-group preferences. They found that the prejudice hypothesis is the most important explanation for negative preferences for neighbourhoods dominated by other ethnic groups. The core of this explanation is that preferences for location are influenced by prejudices about other ethnic groups. Prejudice is understood as being 'heavily imbued with negative affect and negative stereotypes that make the views of the prejudiced group unreceptive to reason and new information'. Bobo and Zubrinski found strong support for this hypothesis among whites. The effects of stereotyping were, however, a good deal less consistent for minority respondents.

Another explanation is rooted in the theory of race prejudice as connected to the social position of ethnic groups (Blumer 1958). People have developed a perception of different ethnic groups as having different social status. It is the magnitude of such differences that in-group members have socially learned to expect and maintain relative to members of specific out- groups that influences their preferences for where to live, especially among the white majority, but also among ethnic minorities. Bobo and Zubrinski examined what they called the *perceived social class difference hypothesis*. It says that negative preferences for living in neighbourhoods dominated by other ethnic minorities are caused by a perception of these neighbourhoods as low-status places. It also implies that 'white' neighbourhoods could be perceived as places dominated by a social class unfamiliar to the one members of an ethnic minority belongs to. It is thus not the ethnic characteristics of a minority neighbourhood that make people abstain from moving to these neighbourhoods but the social content of them. Bobo and Zubrinski did not, however, find strong evidence that perceived differences in status in general played a role in residential integration attitudes.

Ethnic enclaves in Europe

For many reasons the character of ethnic segregation in Europe is different from the US, and it differs greatly between different countries. The question is if ethnic enclaves in the original sense, as defined by Wilson and Portes (1980), are found in Europe besides a few Chinatowns. In many countries, neighbourhoods dominated by ethnic minorities are found in social housing, where the possibilities for doing business are limited. In some countries, like in Britain, such neighbourhoods can be dominated by homeownership or private renting located in older parts of the cities, where ethnic businesses have better conditions. In many places ethnic minorities have started small shops with goods or services that could provide possibilities for employment among newcomers (Bowes et al. 1997 and 2002; Phillips 2006; Phillips et al. 2007). The extent of these local jobs is, however, somewhat limited. Of greater importance could be that living in ethnic enclaves could give access to an ethnic network that has business activities outside the neighbourhood. According to social network theories, the quality of personal contacts is of key importance for getting jobs (Calvó-Armengol and Jackson 2004). This is especially important for ethnic minorities (Damm 2014; Edin et al. 2003). Preferences for living in ethnic enclaves are therefore often connected to expectations of better employment opportunities (Skifter Andersen 2006).

There is a mutual relationship between the character of ethnic concentration and preferences for ethnic enclaves. The existence and intensity of ethnic enclaves depends on the size and composition of ethnic groups and the extent of ethnic segregation in countries and cities. Preferences for ethnic enclaves depend on to what extent local neighbourhoods are available with a sufficient proportion of one's own ethnic group, or other groups with a similar culture, which is decisive for the creation of attractive enclave properties. It is thus important for the existence of enclaves that there is a sufficient number of ethnic minorities coming from the same country or with the same culture. It should also be expected that such preferences are stronger for some ethnic groups than for others. Moreover, it is plausible that ethnic groups with a lower social status, in the country where they reside, have stronger preferences for living separated from the native majority and other groups in accordance with the American hypotheses on the importance of racial prejudices cited earlier. Differences across countries, and between ethnic groups, concerning the social status of ethnic minorities could thus have important influence on ethnic segregation, on the extent of ethnic enclaves and on the importance of preferences for living in neighbourhoods dominated by ethnic minorities.

Formation of enclaves is, however, also dependent on the conditions in the housing market (see Chapters 6 and 7). Creation of enclaves is easier in markets dominated by homeownership where the choice to settle is freer compared to renting, where landlords and social housing companies regulate access. In free markets a dominance of ethnic minorities could lead to lower prices in a neighbourhood because the demand from the native majority could be lower (see Chapter 5), which makes it easier for minorities with lower income to get access.

As described in Chapter 8, the degree of segregation and the character of ethnic neighbourhoods vary greatly across European countries. In most cases ethnic neighbourhoods consist of several ethnic groups. Neighbourhoods dominated by a single ethnic group have a stronger tendency to develop true ethnic enclaves, as defined previously. It should therefore be expected that the existence of enclaves is most prominent in central cities in Britain, Germany and France, seldom in other countries. It is also possible that mixed neighbourhoods dominated by ethnic minorities could develop some of the same advantages for ethnic minorities.

However, as described previously, multi-ethnic neighbourhoods also have a role as a sanctuary for escaping harassment and discrimination from the native majority. It is thus not only the nearness to one's own group that has importance, but that the neighbourhood is not dominated by a native majority with norms and social positions very different from those of the minority groups in question.

Evidence on preferences for neighbourhood among ethnic minorities in Europe

A large number of studies have examined the settlement pattern of ethnic minorities by using quantitative data on what kind of neighbourhood they are living in or moving into. These studies are of little use for an examination of locational preferences, because the housing and location options of ethnic minorities in many ways are limited (see Chapters 6 and 7).

However, a number of studies on stated preferences do exist based on qualitative methods with extensive interviews of a selected number of immigrant groups from a limited number of countries. There are three British studies on Pakistani (Bowes et al. 1997 and 2002; Phillips 2006; Phillips et al. 2007) and one on Bangladeshi (Phillips et al. 2007). There are a number of studies from the Scandinavian countries: Turks, Somali and Polish in Norway (Søholt and Lynnebakke 2015), Somalis and Russians in Finland (Dhalman 2013) and Turks and Somali in Denmark (Skovgaard 2017). Moreover, there is one from the Netherlands on Somali (van Liempt 2011).

A few studies based on surveys have also been conducted. One from the Netherlands on Turks and Surinamese (van der Laan Bouma-Doff 2007) and a Swiss study encompassing all ethnic minorities (Ibraimovic and Masiero 2013). Drever (2004) studied neighbourhood effects in Germany, including households of Turkish, Yugoslavian, Greek, Spanish and Italian origin, where some information can be extracted on neighbourhood preferences. Modood et al. (1997) surveyed all ethnic groups in Britain. Besides these a study has been conducted in Denmark with survey data on the eight largest immigrant groups (Skifter Andersen 2006 and 2015). This study is described in more detail in the last part of the chapter. Only one study, Glikman and Semyonov (2012), is based on survey data from all European countries, which are referenced in Chapter 8.

Most of the qualitative studies found a desire among some members of ethnic minorities for living close by others with the same ethnic background. The intensity of and the reasons for these preferences varied, however, and not all of the interviewed shared the same desire to live close to countrymen. This has also been found inside the same ethnic group, which especially was highlighted by Bowes et al. (2002) in their study of Pakistani in Britain. Preferences for enclaves are primarily found among ethnic minorities coming from outside Europe. Dhalmann (2013) found such preferences among Somalis in Finland, but not among Russians. The study by Søholt and Lynnebakke (2015) showed some preferences among Somalis and Turks in Norway but not among Polish immigrants. The most extensive preferences have been found among South Asian ethnic minorities in Britain and among Somalis in several countries (Dhalman 2013; Søholt and Lynnebakke 2015; van Liempt 2011; and Skovgaard 2017), but also to some extent among Turkish immigrants (Søholt and Lynnebakke 2015; Skovgaard 2017; van Der Laan Bouma-Doff 2007; Drever 2004). Some groups are prepared to renounce housing quality to be able to live in an ethnic neighbourhood. Van Liempt (2011) experienced in some cases Somalis accepting worse housing quality, like overcrowding, if it was located in an area where they could live close together with other Somali families.

The traditional American literature emphasised that especially newly arrived and less integrated/assimilated immigrants should be expected to prefer living in ethnic enclaves. The survey study of van Der Laan Bouma-Doff (2007) showed, however, that acculturation variables, like length of stay in the Netherlands (immigrant status) and command of the Dutch language, seemed to have little impact on preferences. This is quite the opposite of what was found in the Danish study discussed in more detail subsequently (Skifter Andersen 2006 and 2015). The Dutch study showed, however, that more contact with native Dutch led to a lesser probability of choosing co-ethnics as potential neighbours. Phillips (2006) found that especially newly arrived immigrants expressed preferences for living in ethnic neighbourhoods.

Modood et al. (1997) found some differences between different ethnic groups. Among Pakistanis, Caribbeans and Bangladeshi, more than 40 per cent of the respondents preferred to live in neighbourhoods where half or more of the residents belonged to their own ethnic group. For Indians, Africans, Chinese and other Asians, it was only between 25 and 33 per cent.

The survey study in Switzerland conducted by Ibraimovic and Masiero (2013) unexpectedly showed that immigrants from less developed countries showed less preference for ethnic enclaves: 'households belonging to disadvantaged immigrant communities exhibited lower self-segregation preferences compared with the advantaged foreigners'. This is not quite in accordance with what should be expected from other research saying that immigrants coming from developing countries have stronger preferences for ethnic neighbourhoods and to a greater extent tend not to be spatially concentrated. But an explanation for the Swiss results could be that their immigrants have more limited options on the housing market and sometimes are exposed to discrimination. Moreover, there is a large group of Western foreigners in Switzerland that is a somewhat special group of quite rich people. The result was also contradicted by the findings that self-segregation preferences tended to decrease with higher education levels. Households with a lower education level attached more importance to the presence of immigrants from their national background.

The reasons stated for having preferences for enclaves are somewhat similar to the reasons stated in the American literature. For many ethnic groups the feelings of being part of a community, where they can have a rich social life, feel safe and can relax, and where they can easily access ethnic shops and facilities, are important grounds (Phillips et al. 2007; Bowes et al. 2002; Dhalmann 2013). Enclaves are expected to promote a continuing importance of 'community' spaces, which are seen to engender feelings of familiarity, security and support and enhance a sense of belonging and close interconnections. Drever (2004) found, however, in her survey among five ethnic groups in Germany, that ethnic minorities living within ethnic neighbourhoods were no more likely to have a closer connection with the culture of their country of origin than those living elsewhere.

An important reason stated for preferring to live in ethnic neighbourhoods is that it provides access to amenities that facilitate religious and cultural observance. The access to places of worship and to shops delivering ethnic and religious goods, like halal meat, was found important for some individuals (Phillips et al. 2007; Bowes et al. 2002; van Liempt 2011).

However, not all the interviewees in the qualitative studies prefer to live close to ethnic networks. Some want to escape the social control which often is present in ethnic neighbourhoods. Bowes et al. (2002) clarified that there is a variation within the same ethnic group in terms of cultural preferences, suggesting a range of ethnicities within the broad category of 'Pakistani'. Different preferences existed with regard to living close to other Pakistani people, and a minority of respondents actively preferred not to live in a Pakistani area. Van Liempt (2011) stated that it is important to differentiate between those who have actively chosen to live inside the geographical bounds of an ethnic community and those who would actually prefer to live outside. Søholt and Lynnebakke (2015) also found that some Somalis and Turks in Norway preferred not to live near co-ethnics, because of fears of less privacy, social control, gossip or expectations of having an open door, as well as a desire for acculturation and anonymity. This was especially found among Turks who had adapted well to life in Norway and who avoided other Turks following more traditional life-styles, in their neighbourhood.

For some minorities it is not so important to live in a neighbourhood with many co-ethnics in general but that they are close to their tight social network, their family and closer friends. Such preferences are also found among the native majority in countries, but they might be stronger for ethnic minorities. The Canadian study by Fong and Chan (2010) thus showed that a considerable percentage of Asian, Indian and Chinese immigrants in Canada considered proximity to family members to be an important reason for moving into their current neighbourhoods. Dhalmann (2013) also found that 'Even if living near co-ethnic neighbours had lost some of its significance among the Somalis reared in Finland, living near close-kin was still important to them due to the obligations related to the kinship. Family members can meet each other on a daily basis and help each other with different everyday tasks. To facilitate this kind of communal life-style, living in the same neighbourhood or within easy transport connections from parents and siblings was considered desirable by all the interviewees'.

However, preferences for living close to family and friends are often combined with living in ethnic enclaves. Phillips et al. (2007) found that over 70 per cent of households in ethnic neighbourhoods in inner Leeds and Bradford had other people they regarded as family (including close family acquaintances referred to as aunts, uncles, cousins, etc.) living within walking distance. Also, Bowes et al. (2002) uncovered that most of their interviewees had relatives living close by. The survey in Holland by van der Laan Bouma-Doff (2007) showed that respondents' emphasising the importance of having a family network was correlated with preferring co-ethnics as neighbours. This indicates that family ties thus often seem to be related to preferences for ethnic enclaves.

For some minorities a reason for choosing ethnic neighbourhoods is that they find it difficult to socialise with natives, who have another life-style and culture of socialisation (Dhalmann 2013; Søholt and Lynnebakke 2015). In certain contexts self-segregation works as a form of defence in reaction to threats of discrimination and physical harm (Kaplan and Woodhouse 2004). An example is Northern Ireland, where Doherty and Poole (2000) reported that 'segregation levels in Belfast are at an all-time high following a conflict of unprecedented length and intensity'. Some of the qualitative studies found a motive for ethnic enclaves is thus to avoid harassment and discrimination. Bowes et al. (2002) found that harassment had for many been a key impetus in their house moves. Harassment was described as having a serious impact on daily life, restricting movement, threatening people's security and affecting them both physically and mentally. Also Somalis in Finland feared racial harassment and neighbours' prejudiced attitudes (Dhalmann 2013). They saw ethnically mixed areas as safe havens offering a counterforce to racist groups and an opportunity to 'blend in with the crowd'.

In their survey study, van Der Laan Bouma-Doff (2007) found that only minor parts of the ethnic minority groups preferred co-ethnics for imaginary new neighbours, and a large share of respondents answered that they did not have a preference regarding the ethnicity of potential neighbours. A conclusion was that 'measures of interethnic prejudice are the most important predictors of stated preferences for co-ethnic neighbours. These results might reflect the 'safe haven' hypothesis, which states that the choice to live in segregated neighbourhoods is a negative option to avoid hostility and discrimination. The results suggested that preferences for living in ethnic neighbourhoods are driven mostly by interethnic prejudice. In some cases, perceived and experienced hostility and discrimination towards ethnic minorities stimulate self-segregation as well, while interethnic contact decreases it. Also, Glikman and Semyonovs' (2012) study based on European survey data found that perception of discrimination affected immigrants' residential location. Those who claimed that their group members were discriminated against were more likely to dwell in ethnic neighbourhoods. Their data displayed a strong impact of perceived discrimination on residential location, suggesting that those exposed to or worried about discrimination were more likely to live in ethnic neighbourhoods. It was especially among African immigrants that preferences were influenced by fear of discrimination. Asian immigrants were not very much influenced, while immigrants from Muslim countries were influenced to some extent.

It cannot be concluded from this, however, that ethnic minorities want to live in neighbourhoods emptied for natives. Several studies (Bowes et al. 2007; Phillips 2006; Dhalmann 2013; van Der Laan Bouma-Doff 2007; Søholt and Lynnebakke 2015) show that many prefer to live in mixed neighbourhoods composed of both natives and ethnic minorities or that they did not find the ethnic composition important at all. A survey in Britain in the 1990s (Modood et al. 1997) thus reported that more than half of British Pakistanis and Bangladeshis (predominantly Muslim) would be happy to live in areas where their own ethnic group was in the minority. Modood et al. (1997) found some differences across ethnic groups in Britain concerning preferred proportion of ethnic minorities in their neighbourhood (shown in Figure 3.1).

In general, a large proportion did not find the ethnic composition of their neighbourhood important. Pakistanis, Caribbeans and Bangladeshi were most inclined to live in neighbourhoods with a majority of non-native residents. Indians and Africans/Asians mostly did not care about this, while more Chinese wanted a neighbourhood with fewer than half ethnic minorities.

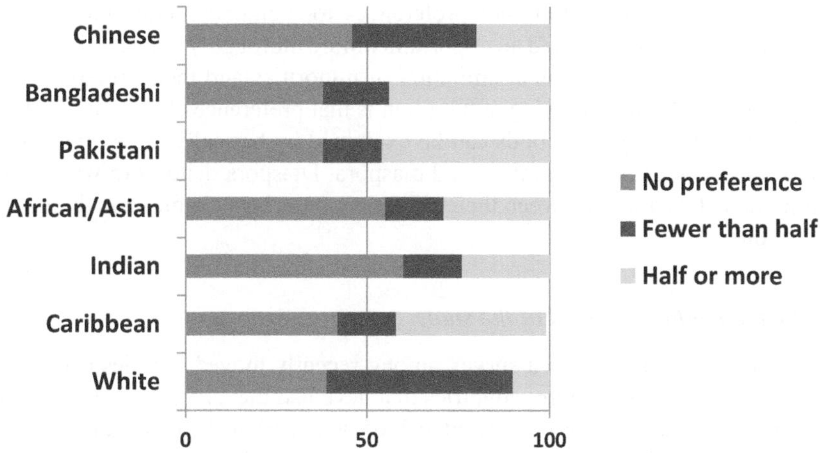

Figure 3.1 Preferred proportion (%) of ethnic minorities in neighbourhood among different ethnic groups in Britain

Source: Modood et al. 1997

Several reasons have been advanced for why ethnic minorities prefer to live in mixed neighbourhoods. Bowes et al. (1997) indicated that some interviewees had the opinion that schools in mixed neighbourhoods were better than in ethnic. Some have also found it important to have good social contacts to natives or other groups to establish a network, learn the language and perhaps most importantly to avoid the social stigmatisation often found in ethnic neighbourhoods and have good neighbours (Phillips 2006; Dhalmann 2013; Bowes et al. 1997).

The connection between neighbourhood preferences and social and economic integration – a Danish study

This section reports the results of a study in Denmark in 2005 based on data from a survey among recently moved immigrants, with information on their housing and settlement preferences. The study focused, among other things, on exposing stated preferences among nine different ethnic groups in Denmark for living in neighbourhoods with access to an ethnic network, and on what explains such preferences, especially how they are connected to social and economic integration. It also highlighted to what extent ethnic minorities want to live in mixed neighbourhoods with native Danes or in neighbourhoods dominated by several ethnic minorities, so-called *multi-ethnic neighbourhoods*. The survey contained data on the social integration among immigrants and on their feeling of diaspora. Moreover, data from public registers on the ethnic composition of their chosen neighbourhood was available (for more detailed information, see Skifter Andersen 2006 and 2015).

The data provided a more reliable basis than earlier research for examining assumptions on what determines preferences for neighbourhood among ethnic minorities. This has provided an opportunity more thoroughly to examine the connection between integration/assimilation of minorities and their preferences for neighbourhood. An underlying assumption is that preferences for ethnic enclaves or multi-ethnic neighbourhoods can be explained by basically three dimensions: their resources, social integration and diaspora. Diaspora denotes to what extent immigrants feel a split between their affiliation to the host country and their country of birth.

Data and methods applied in the study

The data used comes from a survey among recently moved families originating from the nine 'Non-Western' countries that have had the largest immigration into Denmark. The survey was conducted in 2005 and combined with data from public registers on age, incomes, employment and place of birth of persons in every family. Moreover, there is data on their dwelling both before and after the households have moved. All social housing estates are identified in the database, which has made it possible to analyse the composition of residents on the particular social housing estates that ethnic minorities have moved to and from. For more information on the data, see Skifter Andersen 2006 or 2015.

In the analyses of determinants of preferences for ethnic enclaves and for multi-ethnic neighbourhoods, logistic regression was used. A range of background variables like age, family type and degree of urbanisation was used in the analyses. Three different variables were developed as measures of respectively diaspora, social integration and human and economic resources. The measure of diaspora was based on three questions in the survey:

- Do you feel welcome in Denmark?
- Do you wish to go back to your country of origin?
- Do you see television from your country of origin or do you also use Danish media?

The second question especially reveals to what extent the respondents intend to put down roots in Denmark. The answer to the first question can mirror to what extent the respondents have experienced discrimination, which could have consequences for their choice of neighbourhood. The third question has been used in some studies of diaspora (Christiansen 2004) to denote to what extent an immigrant is oriented towards his country of origin. There could be two difficulties in using TV habits in this way. For some ethnic groups it has been difficult to get access to TV from their homeland and in recent years it has been easier to find news on the internet. The last possibility was, however, not so common in 2005 where the survey was completed.

The answers to the survey showed that most of the respondents feel affiliation to Denmark and do not plan to go back to where they came from, and also that quite a few exclusively use TV channels from their country of origin. There is a high correlation between the answers to the first two questions, which points to the fact that the same persons who want to go back also do not feel welcome. The correlation with the use of media is not quite so strong. By means of factor analysis. a measure for diaspora is then constructed. This factor explains 45 per cent of the variation in the variables (for details see Skifter Andersen 2015).

We expected a negative connection between social integration and the preferences for living in multi-ethnic neighbourhoods or enclaves. That is why we chose to construct a variable measuring the opposite of integration called *disintegration*, which could be positively correlated with the measure of diaspora and preferences for enclaves.

In the literature on integration/assimilation of immigrants (see Chapter 4), especially three factors are mentioned as having importance: language skills, participation in the labour market and social connection with the natives. Three questions in the survey have been used to construct the measure of disintegration:

- If the family have Danish friends and how often they are in contact with them
- The ability of the adults in the family to speak and understand Danish
- To what extent members of the family have been in work, and for how long time they have been unemployed

The answers to the three questions were strongly correlated, which meant that it was often the same households that did not speak or understand Danish who did not have Danish friends and who did not have employment. Most of the families claimed, however, that they speak Danish and that they have Danish friends or acquaintances. The proportion with work was somewhat higher than statistical data indicated. In the same way as for the diaspora variable, a variable for disintegration was constructed by factorial analysis of the answers to the three questions.

Resources are of fundamental importance for the options that families have for choosing a home and its location. People's awareness of these options influences settlement preferences. It does not only concern the present income, but also expectations of future incomes, which depend, among others, on education and labour market affiliation. As for the variables on disintegration and diaspora, a composite measure of resources was constructed based on data obtained from public registers on the respondents and their family. The data included:

- The total gross income of the family
- The proportion of incomes coming from employment, not public transfers
- The level of education for the 'main person' in the family, who was chosen as the person with the highest income

A measure for resources was constructed by factor analysis of the three variables. The chosen factor explained 56 per cent of the variation in the variables. It was mostly correlated with the income variables (0.85 and 0.83), but also with the level of education (0.5). As could be expected, this factor has a negative correlation with the disintegration measure (−0.38).

Empirical results

The average factor values for disintegration and diaspora varied greatly across different ethnic groups and there was not a straightforward connection between diaspora and disintegration. For some groups they were connected; for others not. A statistical analysis of which factors could explain the measure of disintegration showed, however, that most of the differences across the groups disappeared when variables like year of immigration, urbanisation of place of origin, age and education were introduced in the analysis (Skifter Andersen 2006). Only four of the nine groups persisted with statistically significant different outcomes, and only two of these, Turks and Somalis, had notable coefficients. The Turks constituted the oldest labour-immigration group and were more integrated and the Somalis less.

Preferences for ethnic enclaves and multi-ethnic neighbourhoods

As discussed earlier, settlement preferences of ethnic minorities can be influenced by preferences for living close to family and friends or to neighbourhoods where they can find an ethnic social network, which can support them in their efforts to be established in their new country. It could be expected that such preferences are correlated with preferences for living in multi-ethnic neighbourhoods where the possibilities of establishing ethnic networks are stronger. But another motive here could also be to avoid Danes and to hide from the demands for cultural integration that immigrants are confronted with in neighbourhoods dominated by Danes.

There are several questions in the survey on preferences concerning the social and ethnic composition of neighbours. The answers from the respondents indicated that living close to countrymen only had minor importance for choosing a neighbourhood, while closeness to family and friends was very important. Thus, it was mostly social networks established before immigrants moved to a certain place that had importance.

Table 3.1 indicates that only a few of the respondents (2 per cent) preferred to live in a multi-ethnic neighbourhood with a majority of immigrants. However, 28 per cent preferred that about half of the residents should be immigrants. Another third preferred that there should be less than one third of immigrants among the residents, while the last third found that this question was without importance for them.

The table shows a clear connection between preferences for multi-ethnic neighbourhoods and the constructed measures for diaspora, disintegration and resources. The few respondents who preferred that more than half of the residents were immigrants had very high scores for diaspora and disintegration and low for resources, while the group that preferred less than one third of immigrants had the highest score on resources and the lowest on diaspora and disintegration.

Thirty per cent of the respondents wanted to live in the same neighbourhood as family and friends or within walking distance. About one third of all respondents

Table 3.1 The distribution of answers to two questions on preferences for ethnic composition of the neighbourhood and on closeness to family and friends, and average measures of resources, disintegration and diaspora

	Distribution of answers %	*Resources*	*Disintegration*	*Diaspora*
What is the most preferable ethnic composition of residents in your neighbourhood?				
Less than one third should be immigrants	28	14	−17	−12
About half should be immigrants	33	−4	8	−6
A majority should be immigrants	2	−37	84	30
The composition does not matter	33	−3	−3	15
Other	3	−46	52	−11
Total	100			
Where do you prefer your family and friends to be located?				
In my own neighbourhood/ housing area	17	−37	60	0
In walking distance	13	−21	35	−5
In short distance with car/bus/train	8	3	−8	7
Somewhere in the town/city	30	17	−25	−18
Without importance	31	9	−14	18
Total	100			

Source: Skifter Andersen 2015

did not find that it was important where friends and family were located, while the remaining third found it sufficient that family and friends were located in the same city. The table very much supports the hypothesis that there is a strong connection between integration and resources and preferences for living close to an ethnic network. Among those who preferred to live in the same neighbourhood or within walking distance there were high scores on disintegration and low on resources. Corresponding to this, there were low scores on disintegration and high scores on resources among those who were satisfied with having family and friends living somewhere in the city or for whom it was without importance.

There was not the same connection with the diaspora variable. The highest score was found among those who did not care where family and friends lived. The reason could be that these were immigrants who did not have family and friends from their native land. Again, the results indicated that preferences do not have a simple connection with characteristics connected with the measure of diaspora.

Figure 3.2 shows for different ethnic groups the proportion who preferred to live in multi-ethnic neighbourhoods, defined as being neighbourhoods with half or more residents belonging to ethnic minorities, and the proportion who wanted to live within walking distance from their family and friends. These preferences differed greatly across different ethnic groups.

As a group, Somalis stand out with their very high preferences for both multi-ethnic neighbourhoods and enclaves. The lowest propensity for living close to family and friends was found among Turks, but at the same time about one third preferred multi-ethnic neighbourhoods. Also Pakistanis, the other big 'labour-immigration' group, had quite high preferences for multi-ethnic neighbourhoods. Among immigrants coming from refugee countries, Iranians had the lowest preferences for living close to family and friends followed by ex-Yugoslavians and Vietnamese. Bosnians had the lowest propensity to move into multi-ethnic neighbourhoods.

Statistical analyses of factors explaining preferences for enclaves and for multi-ethnic neighbourhoods

In the further analyses, two logical variables were used representing respectively preference for enclaves and for multi-ethnic neighbourhoods. The

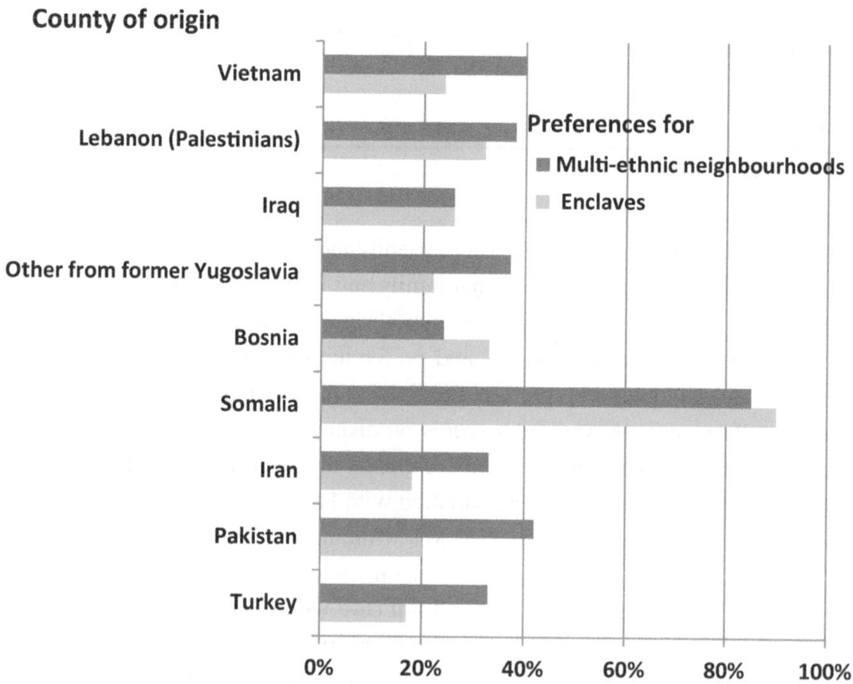

Figure 3.2 Proportion of respondents in different ethnic groups who have preferences for multi-ethnic neighbourhoods and for *enclaves*

Source: Skifter Andersen 2015

variable for preference for enclaves is defined by that the household prefer to live in the same neighbourhood as family and friends or within walking distance, which incorporated on average 30 per cent of the respondents. The variable for preference for multi-ethnic neighbourhoods is defined by households preferring about half of residents, or the majority, should be immigrants. This incorporated 35 per cent.

As shown earlier, there was some connection between ethnic group and the variables disintegration, resources and diaspora. Moreover, other factors might influence residential preferences, for example demographic variables. To determine the isolated importance of each of the factors, a number of logistic regression models were constructed. The first two models concerned preferences for enclaves. In the first model, country of origin was not used, while it was included as a dummy variable in the second analysis. The third and fourth models concerned preferences for living in multi-ethnic neighbourhoods; again, one model without and one with variables for country of origin. The independent variables were:

- Index for diaspora
- Index for disintegration
- Index for resources
- The age of the 'main person', the person in the household with the highest income
- Whether there are children in the household or not
- Whether the family is a couple or not
- Ethnic group (if the household main person belongs to a certain ethnic group or not)

As described here, we expected a positive connection between preferences and diaspora and disintegration. It is not obvious, however, that preferences also depend directly on resources, as resources are correlated with measures of integration. American research has indicated, however, that neighbourhood preferences depend on education, social class and income. Clark (2009) argues that 'Clearly, income does play a role in opening up opportunities and allows higher income individuals to express preferences for neighbourhoods that are less minority and more White'.

We did not have expectations regarding the effects of the demographic variables age, children and couple; they were introduced to control for possible effects of differences in these variables. Moreover, a special variable was introduced in the regressions of preferences for multi-ethnic neighbourhoods: whether the immigrants were living in larger or smaller cities. This variable was included to control for that multi-ethnic neighbourhoods are more seldom in smaller cities and lacking in some of them.

Table 3.2 shows the results of the regressions.

The first two columns in the table indicate that the index of disintegration is a very significant variable explaining preferences for enclaves. But the explanatory power of the first regression (Nagelkerke R^2) is not very strong.

Table 3.2 Results from four logistic regression models explaining preferences for enclaves and for multi-ethnic neighbourhoods

	Preferences for enclaves				*Preferences for multi-ethnic neighbourhoods*			
	Without country variables		*With country variables*		*Without country variables*		*With country variables*	
	Significance	*Odds ratio*	*Significance*	*Odds ratio*	*Significance*	*Odds ratio*	*Significance*	*Odds ratio*
Index for resources	.060	.84*	.689	.96	.625	1.04	.126	1.15
Index for disintegration	.000	1.64**	.008	1.33**	.000	1.37**	.020	1.26**
Index for diaspora	.227	.90	.942	1.01	.243	.91	.564	.95
Age of 'main person'	.292	1.01	.025	1.03**	.183	.99	.660	.99
Have children?	.729	1.06	.573	1.12	.269	1.19	.183	1.25
Couple?	.979	.99	.413	.86	.998	1.00	.628	.93
Living in smaller cities?					.674	.94	.047	.71**
Country of origin								
Turkey			.086	.65*			.760	.94
Pakistan			.730	.85			.043	2.08**
Iran			.077	.51*			.932	1.03
Somalia			.000	48.97**			.000	13.86**
Bosnia			.541	.87			.000	.47**
Other former Yugoslavia			.818	.93			.309	1.30
Iraq			.018	.52**			.103	.68
Lebanon (Palestinians)			.543	1.19			.514	.83
Vietnam			.898	.95			.895	1.04
Nagelkerke R^2	.095		.258		.028		.142	

** Significant at the 5% level. * Significant at the 10% level. (Data Source: Skifter Andersen 2015)

When country of origin is introduced (columns 3 and 4), the regression explains much better the variation in preferences. This indicates that ethnic differences are important in explaining preferences for enclaves. It is, however, seen from the results for the different ethnic groups that only a few of them make a significant difference. The Somali preferences for enclaves are extremely strong. It is not unexpected that they had strong preferences since other research (Kleist 2007) has shown that Somalis have very strong family and clan networks, but it is remarkable that they are so much stronger than for all the other groups. The Somalis only make up about 10 per cent of the survey population, but their divergence is extreme, and this might affect the results. Among the other groups only one is highly significant, Iraqis, and two are somewhat significant, Turks and Iranians. All of them differed by having lower preferences for living close to family and friends.

Table 3.2 also shows that the index for diaspora does not have a significant importance for preferences for enclaves, while the index for resources does not have it for the analysis with country variables. Immigrants with higher resources tend to have lower preferences for enclaves. But the significance disappears when country variables are introduced.

Demographic variables were insignificant except for the age of the main person (with the highest income) of the household. Elderly immigrants have stronger preferences for living close to family and friends than younger ones.

In the right part of the table, the results of the regressions on preferences for multi-ethnic neighbourhoods are shown. In the analysis without country variables, only the index of disintegration is significant, meaning that less integrated immigrants have higher preferences for living in multi-ethnic neighbourhoods with fewer Danes. But again, the explanatory power of the regression is weak. When country variables are introduced, regression results again explain the variation much better. Disintegration is still significant, but with a smaller odds ratio. Moreover, preferences for multi-ethnic neighbourhoods, as expected, are stronger in larger cities.

Again, there is an outstanding result for the Somalis with a high odds ratio; but not as high as for the preferences for enclaves. Only two other groups are significant, but they are not the same groups as the ones found concerning preferences for enclaves. Pakistanis have a relatively strong preference for multi-ethnic neighbourhoods, while Bosnians have weaker preferences than other groups.

The results suggest that there is not a straightforward connection between preferences for living close to family and friends and for multi-ethnic neighbourhoods among different groups. Only for the Somalis both kind of preferences are very strong. An interpretation could be that for some groups, the disadvantages of living in multi-ethnic neighbourhoods have greater importance, which could counteract their wish to live close to family and friends. They could have a strategy to establish an ethnic network in more mixed urban areas. This does not, however, explain the preferences of the Pakistanis, which seem to be the opposite.

Implications of the Danish study

The referenced study from Denmark showed little inclination among Non-Western minorities to live in neighbourhoods with a majority of immigrants, while about one third preferred an equal mix of Danes and minorities, which indicates that the sanctuary motive has some importance.

For some minorities it is not so important to live in ethnic enclaves, but it is more important that they are close to their tight social network: their family and closer friends. The Danish study showed that this preference had a strong connection with variables expressing the social and economic integration, like language proficiency, social contacts with Danes, employment, income and education. Preferences for living in neighbourhoods close to co-ethnics are thus very much determined by the degree to which immigrants have not been able to be an integrated part of their new society. The study showed that on average preferences varied greatly between the nine ethnic groups included in the study, but statistical analysis found that when controlling for social integration and resources, the differences across the groups mainly disappeared. There were, however, some significant differences between a few of the immigrant groups, which cannot be explained by their poor measure of integration. A possible explanation could be that a greater feeling of discrimination and exclusion among some groups lead to the sanctuary motive gaining more importance. However, cultural explanations were also advanced. Some immigrants continue to have a strong affiliation to their country of origin and feel a split between their belonging to their new and old country, described by the concept of diaspora. This could impact their inclination to settle in enclaves or multi-ethnic neighbourhoods. Some ethnic groups could also have a culture that promotes a stronger wish to live in enclaves to maintain a separate cultural identity.

References

Abrahamson M. (2004) *Global Cities*. Oxford: Oxford University Press.

Abrahamson M. (1995) *Urban Enclaves: Identity and Place in America*. New York: St. Martin's Press.

Ærø T. (2006) Residential Choice from a Lifestyle Perspective, *Housing, Theory and Society*, 23, pp. 109–130.

Alba R. and Nee V. (1997) Rethinking Assimilation Theory for a New Era of Immigration, *International Migration Review*, 31(4), pp. 826–874.

Barabantseva E. (2016) Seeing Beyond an 'Ethnic Enclave': The Time/Space of Manchester Chinatown, *Identities*, 23(1), pp. 99–115.

Blumer H. (1958) Race Prejudice as a Sense of Group Position, *Pacific Sociological Review*, 1, pp. 3–7.

Bobo L. and Zubrinski C. L. (1996) Attitudes on Residential Integration. Perceived Status Differences, Mere In-Group Preference or Racial Prejudice? *Social Forces*, 74(3), pp. 883–909.

Bowes A., Dar N. and Sim D. (1997) Tenure Preference and Housing Strategy: An Exploration of Pakistani Experiences, *Housing Studies*, 12(1), pp. 63–84.

Bowes A., Dar N. and Sim D. (2002) Differentiation in Housing Careers: The Case of Pakistanis in the UK, *Housing Studies*, 17(3), pp. 381–399.

Bushi M. (2014) Rethinking Heterolocalism: *The Case of Place-Making among Albanian-Americans*, Geography Honors Projects. Paper 40.

Calvó-Armengol A. and Jackson M. O. (2004) The Effects of Social Networks on Employment and Inequality, *American Economic Review*, 94(3), pp. 426–454.

Christiansen C. C. (2004) News Media Consumption among Immigrants in Europe: The Relevance of Diaspora, *Ethnicities*, 4(2), pp. 185–207.

Clark W. A. V. (1992) Residential Preferences and Residential Choices in a Multiethnic Context, *Demography*, 29, pp. 451–466.

Damm A. P. (2014) Neighborhood Quality and Labor Market Outcomes: Evidence from a Quasi-random Neighborhood Assignment of Immigrants, *Journal of Urban Economics*, 79, pp. 139–166.

Dhalmann H. (2013) Explaining Ethnic Residential Preferences – The Case of Somalis and Russians in the Helsinki Metropolitan Area, *Housing Studies*, 28(3) pp. 389–408.

Doherty P. and Poole M. (2000) Living Apart in Belfast: Residential Segregation in a Context of Ethnic Conflict. In Boal F. *Ethnicity and Housing: Accommodation Differences*, pp. 179–189. Brookfield, VT: Ashgate.

Drever A. I. (2004) Separate Spaces, Separate Outcomes? Neighbourhood Impacts on Minorities in Germany, *Urban Studies*, 41(8), pp. 1423–1439.

Edin P.-A., Frederiksson P. and Åslund O. (2003) Ethnic Enclaves and the Economic Success of Immigrants – Evidence From a Natural Experiment, *The Quarterly Journal of Economics*, 118(1), pp. 329–357.

Farley R., Fielding E. L. and Krysan M. (1997) The Residential Preferences of Blacks and Whites: A Four-metropolis Analysis, *Housing Policy Debate*, 8(4), pp. 763–800.

Fong E. and Chan E. (2010) The Effect of Economic Standing, Individual Preferences, and Coethnic Ethnic Resources on Immigrant Residential Clustering, *International Migration Review*, 44(1), pp. 111–141.

Glikman A. and Semyonov M. (2012) Ethnic Origin and Residential Attainment of Immigrants in European Countries, *City & Community*, 11(2), pp. 198–219.

Hatziprokopiou P. and Montagna N. (2012) Contested Chinatown: Chinese Migrants' Incorporation and the Urban Space in London and Milan, *Ethnicities*, 12(6), pp. 706–729.

Ibraimovic T. and Masiero L. (2013) Do Birds of a Feather Flock Together? The Impact of Ethnic Segregation Preferences on the Neighbourhood Choice, *Urban Studies*, 2013, pp. 1–19.

Johnston R., Forrest J. and Poulsen M. (2002) Are there Ethnic Enclaves in English Cities? *Urban Studies*, 39, pp. 591–618.

Kaplan D. H. and Woodhouse K. (2004) Research in Ethnic Segregation I: Causal Factors, *Urban Geography*, 25(6), pp. 579–585.

Kinkead G. (1992) *Chinatown: A Portrait of a Closed Society*. New York: HarperCollins.

Kleist N. (2007) *Spaces of Recognition: An Analysis of Somali-Danish Engagement and Diasporic Mobilization*. PhD thesis. Institute of Sociology, Copenhagen: University of Copenhagen.

Logan J. R., Zhang W. W. and Alba R. D. (2002) Immigrant Enclaves and Ethnic Communities in New York and Los Angeles, *American Sociological Review*, 67(2), pp. 299–322.

Luk W. K. (2008) *Chinatown in Britain: Diffusions and Concentrations of the British New Wave Chinese Immigration*. Amherst, MA: Cambria Press.

Malheiros J. (2002) Ethni-cities: Residential Patterns in the Northern European and Mediterranean Metropolises – Implications for Policy Design, *International Journal of Population Geography*, 8, pp. 107–134.

Marcuse P. (1997) The Enclave, the Citadel, and the Ghetto. What has Changed in the Post-Fordist U.S. City, *Urban Affairs Review*, 33, pp. 228–244.

Massey D. S. (1985) Ethnic Residential Segregation: A Theoretical Synthesis and Empirical Review, *Sociology and Social Research*, 69, pp. 315–350.

Massey D. S. and Denton N. A. (1993) *American Apartheid: Segregation and the Making of the Underclass*. New Haven, CT: Harvard University Press.

Massey D. S. and Denton N. A. (1985) Spatial Assimilation as a Socioeconomic Outcome, *American Sociological Review*, 50(1), pp. 94–106.

Modood T., Berthoud R., Lakey J., Nazroo J., Smith P., Virdee S. and Beishon S. (1997) *Ethnic Minorities in Britain: Diversity and Disadvantage*. London: Policy Studies Institute.

Park R. E. (1925) The City: Suggestions for the Investigation of Human Behaviour in the Urban Environment. In Park R. E. and Burgess E. W. *The City*, pp. 1–46. Chicago, IL: University of Chicago Press.

Park R. E. and Burgess E. (1969) *Introduction to the Science of Sociology*. 1921. Reprint. Chicago, IL: University of Chicago Press.

Peach C. (1998) South Asian and Caribbean Ethnic Minority Housing Choice in Britain, *Urban Studies*, 35(10), pp. 1657–1680.

Phillips D. (2009) Minority Ethnic Segregation, Integration and Citizenship: A European Perspective, *Journal of Ethnic and Migration Studies*, 36(2), pp. 209–225.

Phillips D. (2006) Parallel Lives? Challenging Discourses of British Muslim Self-segregation, *Environment and Planning D: Society and Space*, 24, pp. 25–40.

Phillips D., Davis C. and Ratcliffe P. (2007) British Asian Narratives of Urban Space, *Transactions of the Institute of British Geographers*, 32(2), pp. 217–234.

Portes A. and Manning R. (1986) The Immigrant Enclave: Theory and Empirical Examples. In Nagel J. and Olzak S. *Competitive Ethnic Relations*. Orlando, FL: Academic Press.

Qadeer M. and Kumar S. (2006) Ethnic Enclaves and Social Cohesion, *Canadian Journal of Urban Research*, 15(2), pp. 1–17.

Skifter Andersen H. (2015) Explanations for Special Neighbourhood Preferences among Ethnic Minorities, *Housing, Theory and Society*, 32(2), pp. 196–217.

Skifter Andersen H. (2011) Explanations for Preferences for Surroundings and Location of the Home, *Urbani izziv*, 22(1), pp. 100–114.

Skifter Andersen H. (2010) Spatial Assimilation in Denmark. Why do Immigrants move to and from Multi-ethnic Neighbourhoods? *Housing Studies*, 25(3), pp. 281–300.

Skifter Andersen H. (2006) *Bo sammen eller spredt? Etniske minoriteters boligønsker og motiver for bosætning*. Horsholm: Danish Building Research Institute.

Skovgaard Nielsen R. (2017) The Potentials of a Strong Social Housing Sector: The Case of Turks and Somalis in the Copenhagen Housing Market, *Housing, Theory and Society*, 34(4), pp. 458–476.

Søholt S. and Lynnebakke B. (2015) Do Immigrants' Preferences for Neighbourhood Qualities Contribute to Segregation? The Case of Oslo, *Journal of Ethnic and Migration Studies*, 41(14), pp. 2314–2335.

South S. J., Crowder K. and Chavez E. (2005) Migration and Spatial Assimilation among U.S. Latinos: Classical Versus Segmented Trajectories, *Demography*, 42(3), pp. 497–521.

van der Laan Bouma-Doff W. (2007) Involuntary Isolation: Ethnic Preferences and Residential Segregation, *Journal of Urban Affairs*, 29(3), pp. 289–309.

van Liempt I. (2011) From Dutch Dispersal to Ethnic Enclaves in the UK: The Relationship between Segregation and Integration Examined through the Eyes of Somalis, *Urban Studies*, 48(16), pp. 3385–3398.

Wacquant L. (1999) Urban Marginality in the Coming Millennium, *Urban Studies*, 36(10), pp. 1639–1647.

Waldinger R. (1993) The Ethnic Enclave Debate Revisited, *International Journal of Urban and Regional Research*, 17(3), pp. 444–452.

Wilson K. and Portes A. (1980) Immigrant Enclaves: An Analysis of the Labor Market Experiences of Cubans in Miami, *American Journal of Sociology*, 88, pp. 295–319.

Wright R., Ellis M. and Parks V. (2005) Re-placing Whiteness in Spatial Assimilation Research, *City and Community*, 5(2), pp. 111–135.

Zelinsky W. and Lee B. A. (1998) Heterolocalism: An Alternative Model of the Sociospatial Behaviour of Immigrant Ethnic Communities, *International Journal of Population Geography*, 4, pp. 281–298.

Zhou M. (1992) *Chinatown: The Socioeconomic Potential of an Urban Enclave*. Philadelphia, PA: Temple University Press.

4 Spatial assimilation – the development in immigrants' locational preferences and choices with duration of stay

Ethnic segregation is not a static phenomenon. People move to and from neighbourhoods all the time, changing the ethnic and social content of them. This is to some extent connected to the changes that occur over time for individual members of ethnic minority groups. As natives do, they experience changes in their housing needs and preferences during their life time that can be attributed to changes in their family situation and work career. The special condition for immigrants is, however, that at some point in time they, or their ancestors, did arrive to a new country where many of them had poorer prerequisites for making a good existence, mostly because of language problems and poor education, but also because of inadequate knowledge about important norms and rules and because of lack of social connections to important people and institutions. Over the course of time, sometimes generations, most immigrants have succeeded in overcoming such barriers and improved their skills for managing a better life. The most common terms used for such a development is that they have been integrated or assimilated into their new society.

The mode of arrival to the country, and the way they settled in the first place, have had substantial importance for their possibilities for having success and for their choice of housing and neighbourhood. Some immigrants came as a result of what has been called chain migration, sometimes by family reunification or by having friends and family who already had become established themselves. Such immigrants have had better conditions for doing well because of the support they could get from their network. Often, they also got help to find housing, sometimes residing with family members. Other immigrants, who came as refugees or had no network to support them, have faced a much more difficult position.

The historic mode of immigration might have had a special importance for the character and extent of ethnic segregation. If earlier arrived immigrants have settled in multi-ethnic neighbourhoods, chain migration has contributed to further segregation when new immigrants have settled close to their network. For other kinds of new immigrants, it may have been more difficult to find housing in ethnic neighbourhoods depending on the character of these places. The initial settlement of immigrants and their later residential career have depended much on the character of the housing market and its spatial structure, as will be discussed in Chapters 6 and 7. If ethnic neighbourhoods were slum areas with low quality and cheap housing, they might have been easy to access for all new immigrants, but if they were more attractive neighbourhoods, like many social housing estates in Europe, access may have been more difficult just after arrival. Moreover, in many countries governments have imposed special rules for how to settle refugees, trying to spread them across the country and across different neighbourhoods.

The dominant theory on how immigrants settle, when they arrive and how they change residence over time, including following generations, is called the *spatial assimilation theory*. It originated from the theory developed by the American 'ecological' theoretical tradition of the Chicago School (Park 1928; Park and Burgess 1969). The spatial location of immigrants was assumed to be closely connected with their social, cultural and economic assimilation in the host society. This theory claimed that newly arrived immigrants will tend to cluster in certain neighbourhoods, dominated by their own ethnic group, and that they in the course of time will be more dispersed in the city in pace with their integration/assimilation. This dispersal can be unleashed by two different reasons. The first is that immigrants over the course of time will improve their economic situation, which creates a demand for better housing not found in the ethnic neighbourhoods. The other is that immigrants' preferences for living in ethnic enclaves decrease in pace with their social and cultural integration.

The spatial assimilation theory is based on a chain of assumptions:

1 That newly arrived immigrants prefer to settle in ethnic enclaves, which often are found in immigrant-dense neighbourhoods
2 That over the course of time they will be culturally, socially and economically integrated ('assimilated' in the US literature) in their new country
3 That this integration will change their preferences for location in ethnic enclaves, and
4 That these changed preferences will result in migration away from neighbourhoods with high concentrations of ethnic minorities to neighbourhoods with better housing and environment, dominated by the native population.

All these assumptions will be discussed in this chapter. In the research literature several of them have been doubted depending on which ethnic group or country has been studied.

Spatial assimilation can in principle be examined in different ways. The best method is to follow immigrants' settlement pattern over time and explore the importance of aspects of integration for changes in their choice of neighbourhood, but such studies demand data that are difficult to provide. Some studies examine how ethnic minorities move within a fixed period of time and test if there is a connection between their integration and to what extent they move from ethnic neighbourhoods to other neighbourhoods. The last method is to examine the movements in and out of ethnic enclaves or multi-ethnic neighbourhoods to study if the pattern corresponds to what is expected from the spatial assimilation theory, saying that in-movers are recent and less integrated immigrants while out-movers have been longer in the country and are better integrated. These three kinds of relatively few studies on spatial assimilation in Europe will be reviewed in this chapter. Moreover, results from two recent Danish studies will be used to illuminate the issue in more detail. The first is a study of in- and out-movers from multi-ethnic neighbourhoods examining differences in their characteristics, especially concerning aspects of integration. The other is a longitudinal study of immigrants' settlement patterns over the course of time after their arrival in the years from 1985 to 2008.

Research on the basic assumptions of the theory

The origins of the spatial assimilation theory

The spatial assimilation theory arose as a result of the research conducted by the so-called Chicago School in the beginning of the 20th century on how immigrants to the US over time adapted to their new society (Park 1928; Park and Burgess 1969). In the literature on the development of immigrants' situation over time, an assimilation perspective has hereafter dominated much of the sociological thinking on the subject. Central to this perspective are the assumptions that there is a natural process by which diverse ethnic groups come to share a common culture and to gain equal access to the opportunity structure of society; that this process consists of gradually deserting old cultural and behavioural patterns in favour of new ones; and that, once set in motion, this process moves inevitably and irreversibly towards what is called integration/assimilation. Classical assimilationists argued that migration leads to a situation in which immigrants are pulled in the direction of the host culture but drawn back by the culture of their origin (Park 1928; Alba and Logan 1992; Zhou 1997). As a consequence of this assimilation to the host culture, it was assumed that immigrants over time also were dispersed spatially in the American city, living side by side with the majority white population.

The spatial assimilation theory postulated that immigrants often started their career in the US by moving to neighbourhoods with many residents from their home country, but that most of them after some time moved out again to housing and neighbourhoods that were more in accordance with changes in their preferences, resources and needs. Massey (1985) formulated a spatial assimilation model, whose propositions have generally been supported empirically in a large number of American studies (Alba and Logan 1992).

The Chicago School's ecological model was formulated at a time when immigration was dominated by low-wage European labour. Newcoming immigrants were poor and were expected to settle in the worst slums in the cities where they could find cheap, but low quality, housing. In his model, Massey used elements from the ecological model of ethnic succession in neighbourhoods, hypothesising that ethnic enclaves are located in older, central city neighbourhoods. Preferences among ethnic minorities for moving to and from ethnic neighbourhoods were assumed to depend on the extent to which they were socially and economically integrated in the new society. In pace with their economic and social integration, immigrants were expected to move out from the slums to neighbourhoods with better housing and neighbourhood amenities (e.g. good schools, clean streets). Especially since the 1950s, such neighbourhoods in the US have been largely suburban (Alba and Logan 1992). Spatial assimilation in the US after World War II has therefore been connected to migration from central cities to suburbs.

Over the past 35 to 40 years, the composition of recent immigrants has changed substantially in Canada and the US (Haan 2007), resulting in an immigrant population that is much more racially diverse than it was in the past. This change has spurred several scholars to question the ability of traditional understandings of assimilation to account for the processes of integration into the mainstream of their new societies and their spatial integration. The assumption about new immigrants' settlement in central city enclaves has also been questioned (see Chapter 3). It has been claimed that they now settle more dispersed in the city from the beginning, forming ethnic networks across neighbourhoods, so-called heterolocalism (Zelinsky and Lee 1998). It has also been found that new immigrants sometimes move into ethnic enclaves that are located in suburbs (Alba et al. 1999; Clark and Patel 2004; Wright et al. 2005).

Studies on spatial assimilation have been much sparser in Europe. Moreover, the relationship between social integration and spatial segregation continues to be contested, especially in countries where the level of segregation is low and where welfare policies promoting social equality are strong (Wessel et al. 2017; Bolt et al. 2010; Bolt and Van Kempen 2010).

The settlement pattern of newly arrived immigrants

A basic assumption in the spatial assimilation theory is that a driving force in the spatial segregation of ethnic minorities is that new immigrants tend to settle in neighbourhoods dominated by their own group. Newly arrived immigrants have mostly been poor and had poorer prerequisites for making a good existence, because of language problems and poor education, but also because of poor knowledge about important norms and rules and lack of social connections to important people and institutions. As described in Chapter 3, it has been shown that some members of ethnic minority groups for different reasons have preferences for living in neighbourhoods with an ethnic social network and access to ethnic services and goods; sometimes also with better possibilities to get work through contacts to minority employers. It was shown that these preferences were stronger among those who were less socially integrated and who had smaller incomes and less education. It is thus to be expected that newly arrived immigrants will have stronger incentives to move into ethnic enclaves or multi-ethnic neighbourhoods.

A study in some middle-sized cities in Canada (Texeira and Drolet 2016) indicated that most of the new immigrants had social contacts in the cities, friends, relatives or employers, before they arrived. There were no well-established ethnic communities of compatriots in these cities to help them settle, but the network of contacts they already had established played a determining role in helping the majority of new immigrants to find a place to live and/or a job when they first arrived. These findings were congruent with those of other Canadian studies in small- and mid-sized cities, as well as major Canadian cities, which show that new immigrants' heavy reliance on their own (informal) social networks represents a very important coping strategy in adapting to a city's social, cultural and economic environment.

It has been assumed that the kind of neighbourhoods that new immigrants settle in are what have been called *immigrant enclaves* (see Chapter 3). A study in greater Los Angeles (Wright et al. 2005) thus found that newly arrived immigrants were more spatially concentrated than established immigrants, although the effect varied by group. In the earlier American literature on the settlement of European immigrants, immigrant enclaves have mostly been confined to disadvantaged neighbourhoods with poor housing and low-income residents in central cities (Myles and Hou 2003). Newer studies, however, indicate a quite different pattern of settlement. It has been found that some new immigrants frequently settle in suburbs immediately upon, or soon after, arrival in the United States (Alba and Logan 1992; Alba et al. 1999). This hallmark of new immigration presents a remarkable contrast to the process of spatial assimilation from central cities to suburbs as experienced by earlier, European, immigrant groups. Clark and Patel (2004) studied how recent immigrants in Los Angeles spread in the city and found that the central city still continued to receive lower income immigrants, with lower levels of human capital, but that also the suburbs received both households with lower and higher levels of human capital. They concluded that the spatial patterns of new immigrants have become more complex than in the past. The central city suburban dichotomy, while still relevant, may not be the best way to analyse the settlement pattern over time of immigrants in the US.

Other scholars have questioned the basic assumption that newly arrived immigrants have to settle in ethnic neighbourhoods to obtain a social network. Zelinsky and Lee (1998) proposed a theory called *heterolocalism*, saying that ethnic networks do not necessarily have to be established in a certain geographical area, but can exist across different neighbourhoods. They proposed that, just after arrival, there might be an immediate or prompt spatial dispersion of new heterolocal immigrants within the host country (see also Chapter 3).

The settlement pattern of new immigrants in Europe depends much on the kind of immigrants the countries receive; labour immigration, refugees or chain migration/family reunification; or if they come from poor or rich countries. Immigrants that come by family reunification often are housed upon arrival by the family, most often own or spouse's parents. A Dutch study found that this differed greatly between two different ethnic groups from Morocco and Turkey (Bolt and Van Kempen 2002), while the Danish study described subsequently did not find notable differences between immigrants from different parts of the world. Moreover, the conditions in the receiving country, as well as the extent of immigration policies, housing market conditions and to what extent immigrants get help from local authorities to obtain housing, have importance.

The general policy in countries for how to receive *refugees* has significant importance for how and where they settle. In some countries, refugees get help from local authorities to get housing or are subject to a compulsory allocation of where to live. An analysis by Scholten et al. (2017), comparing ten European countries, concluded that there were significant differences between the countries concerning assistance to refugees to obtain housing. Approximately half of the countries grant refugees support to find a permanent accommodation, while the other half leaves this process to the full responsibility of the recognised individuals. However, in most countries at least financial support is granted for independent housing. In some countries (Denmark, Norway and Germany) refugees are allocated to municipalities according to fixed rules of geographical distribution. The local authorities then decide where they can settle inside the municipality. Sometimes local authorities use these powers to prevent that refugees settle in ethnic neighbourhoods and are spread in the city and countryside. In some cases, refugees are bound to remain in the same municipality for a certain period.

New immigrants from third world countries usually are quite poor and have difficulties in getting well-payed jobs. Some of them can get welfare benefits, but in some countries like Britain and Denmark there are restrictions or reduced benefits. Their options on the housing market are very limited and some of them cannot even afford to live in their own dwelling, but have to live with friends or family or to stay in dormitories or places lacking facilities.

According to an extensive study in the UK (Robinson et al. 2007), the arrival experiences of new immigrants were largely determined by their immigration status, associated legal rights and related opportunities. This led to clear distinctions in the housing situations of different groups. For example, asylum seekers, once granted leave to remain, were initially reliant on the National Asylum Support Service for accommodation and subsequently moved into social housing. In contrast, the restricted rights of migrant workers to welfare benefits meant they were reliant on the private rented sector for a place to live. New immigrants tended to fill voids in the housing stock left behind or avoided by other households. The result was the concentration of new immigrants in particular sectors of the local housing market and in specific neighbourhoods. Such dwellings are often located in older dilapidated neighbourhoods or in newer deprived estates in the suburbs.

Most new immigrants moved into temporary accommodation upon first arrival in the UK. Poor living conditions, lack of privacy and concerns about safety and security were often associated with temporary accommodation and were sometimes endured for many months (Phillips 2007). The new immigrants did not live in isolated ethnic clusters and their residential settlement patterns were rarely the outcome of self-segregating tendencies. They quickly developed an affiliation with the place where they lived, which resulted in many wanting to stay in the neighbourhood where they first settled.

A study (Robinson 2008) in Britain showed that relatively few new immigrants got access to social housing. In many other countries social housing is an important sector for them, which often means better housing possibilities. In the Netherlands, almost all refugees in 2014 were thus housed by social landlords (www.

housingeurope.eu/blog-514/housing-refugees-in-the-netherlands). The next section shows that on average up to 40 per cent of new immigrants coming to Denmark in the last 25 years started their housing career in social housing. For immigrants who were family reunified with an earlier immigrant nearby, 60 per cent settled in social housing. Of decisive importance for the spatial location of new immigrants is then to what extent they are allocated to housing estates with a high or low concentration of immigrants and to what extent social housing is dispersed or concentrated in certain neighbourhoods in the cities, as discussed in Chapter 7. The study in Denmark showed that only 18 per cent of newly arrived immigrants in the last 25 years have settled in neighbourhoods with more than 20 per cent Non-Western immigrants.

Zorlu and Mulder (2008) conducted an in-depth analysis of how new immigrants who came to the Netherlands in 1999 settled. The results confirmed that their location choices were largely determined by if the neighbourhood contained ethnic minorities from the same origin. There were, however, significant differences between Non-Western ethnic groups. Some groups (Turks and Moroccans) more often moved to a neighbourhood where the proportion of homeownership was relatively high, however mostly with cheap housing. They also found differences between immigrants having different reasons for residence permits in the Netherlands. *Labour migrants*, frequently coming from Western countries, chose neighbourhoods with higher levels of earnings. The settlement pattern of *asylum migrants*, who were subject to government settlement policies, was strongly influenced by the share of other Non-Western immigrants. Moreover, the location was apparently not influenced by good economic prospects: asylum migrants were more frequently located in municipalities with higher unemployment rates. Those who entered the Netherlands on the basis of *family reunification* seemed to be slightly less sensitive to both the presence of co-ethnics and the economic indicators of the residential location (mean wage and unemployment rate). It was concluded that this outcome did not conform to the expectation that immigrants with a family reunification motive settle more frequently in neighbourhoods where their co-ethnics are concentrated. This is in contrast to the results from the Danish study reviewed in the last section.

Immigrants integration/assimilation over time

For two reasons the integration of immigrants in their new societies influences their spatial location and the extent to which they live separated from the native majority or live in mixed neighbourhoods. The first concerns their social and cultural integration, which might influence their propensity to prefer to have neighbours from their own ethnic group and to live in ethnic enclaves. The other is their integration in the economic system, which determines their financial situation and their chances to settle in more prosperous neighbourhoods together with the native majority.

In the classical literature (Gordon 1964) on immigration and social integration in the United States and Canada, the theory was proposed that immigrants and their descendants during the course of time are *assimilated* into the host society sharing fundamental norms and values of the native population.

Gordon (1964) distinguished between *acculturation* and what he termed *structural assimilation*. Acculturation refers to the minority group's adoption of the *cultural patterns* of the host society. Most important was here the acquisition of the English language. By structural assimilation he meant the entry of members of an ethnic minority into primary group relationships with the majority group. This especially concerns the extent to which immigrants are participating in the labour market and have social relations to natives and other groups in society.

Gordon also defined *straight-line assimilation*, a process unfolding in a sequence of generational steps; each new generation represents on average a new stage of adjustment to the host society. The idea of the generational inevitability of assimilation has been criticised, however, for assuming that all ethnic content is imported by immigrants and not recognising that it can be created in response to conditions and out of cultural materials in the host society. As stated by Alba and Nee (1997): 'Over time ethnic groups were expected to become assimilated into their host society: most did economically at least, but not always socially' and 'assimilation involves the decline, though not always the disappearance of ethnic/racial distinctions'.

The idea of straight-line assimilation has been questioned in relation to the European experience and it has also been questioned in connection with new waves of ethnic groups coming to the US and Canada. Over the past 35–40 years, the composition of recent immigrants has changed substantially, resulting in an immigrant population that is much more racially diverse than it was in the past. This change has spurred several scholars to question the ability of traditional understandings of immigrant assimilation to account for the processes of immigrant integration into the mainstream of their new societies. Critics of the straight-line notion have argued that, instead, ethnicity may go through periods of recreation, if not renaissance (Glazer and Moynihan 1970; Conzen et al. 1992). Others maintain that the theory of assimilation still holds in the US, but that it takes more time for the newer waves of immigrants, sometimes several generations (Alba and Nee 1997).

Another point of criticism, according to Alba and Nee (1997), is that occupational mobility and economic assimilation, the key dimensions of socio-economic integration, are not properly addressed in Gordon's discussion of assimilation. Yet this kind of assimilation is of paramount significance because parity of life-chances with natives is a critical indicator of the decline of ethnic boundaries. Socio-economic mobility creates the social conditions conducive to other forms of assimilation since it likely results in a more equal status and in more contact across ethnic lines in workplaces and neighbourhoods. Socio-economic assimilation can be defined as minority participation in institutions such as the labour market and education on the basis of parity with native groups.

For some immigrants it can be important to sustain cultural connections with the home country through familiar language, social networks and ethnic organisations (Wright et al. 2005). This motive is connected with the concept of *diaspora*, which denotes to which extent immigrants feel a split between their affiliation to the host country and their country of origin. Some talk about *transnationalism*, meaning that immigrants and their descendants continue to have an international network upon which they can depend.

Diaspora can according to Clifford (1999) be described as a 'continuing relationship to the homeland'. Originally, diaspora indicated the dispersal of a population that maintained their ethnic affiliation, especially the dispersal of Jews. Diaspora identity is characterised by a tension between relations or belonging to some kind of original home or nation, now left behind, and to the place where one has now settled and the larger, usually ethnic, community found there (Clifford 1999; Cohen 1997; Gilroy 1996).

A problem with classical American assimilation theory is that it predicts a gradual convergence to the socio-economic outcomes of the white middle class, called Anglo-conformity. American culture varies, however, greatly by locality and social class. It is argued (Alba and Nee 1997) that classical American theory made a mistake in presupposing integration into the values of the white middle class. Some researchers have formulated a theory of *segmented assimilation*, where immigrants and subsequent generations are integrated into different segments of American culture and some into 'permanent poverty and assimilation into the underclass' (Valdez 2006; Portes and Rumbaut 2001), what has been coined *downward assimilation* (Model 1991; Portes and Zhou 1993). Several scholars challenge the prospect of equal opportunity as a universal end point, arguing instead that no matter how long the time frame, some groups will never fully converge upon the native-born (Lee and Bean 2004). Portes and Zhou (1993) found that some groups, like the Haitians, appear to enter into a pre-existing race relations cycle, and are being pressured into a racially delineated underclass. Rather than follow a gradual and progressive journey into the mainstream, some immigrants are encountering similar obstacles to those faced by African-Americans (Waters 1999; Alba et al. 2003; Lee and Bean 2004), facing segmented assimilation. Others have argued that segmented assimilation does not necessary imply an assimilation into the underclass (Portes and Zhou 1993; Zhou 1997; Greenman and Xie 2008). They emphasise that there is more than one way of 'becoming American', and that Americanisation is not necessarily always beneficial because of racial conflicts. There is a possibility that deliberate preservation of the immigrant community's culture and values can be accompanied by economic integration.

According to other scholars, classical assimilation theory, in its original form, is still relevant (Brubaker 2001; Nagel 2009; Alba and Nee 1997; Greenman and Xie 2008; Waldinger and Feliciano 2004). They argue that it is not clear that differences between current and past immigrants are significant enough to render classical assimilation theory inapplicable. It has been contended that the experience of today's immigrants and their offspring is not truly all that different from that of the 1890–1920 wave of European immigrants. For example, Alba and Nee (1997, 2003) argue that the offspring of earlier European immigrant groups often did not fully assimilate until the third or fourth generation. Thus, observations of limited assimilation among today's second-generation youth should not be surprising. Waldinger and Feliciano (2004) argue that Mexican immigrants, who are often considered the group most vulnerable to downward assimilation, transition into the American working class in a similar manner as earlier large immigrant groups. They show that their labour force outcomes appear to be converging across generations with native whites rather than with native minority groups.

Results from more recent studies of the development in employment and education for immigrants and their descendants in the US and Canada show signs of an assimilation over the course of time, but there are big differences between different groups (Valdez 2006; Greenman and Xie 2008; Slack and Jensen 2007; Haan 2007). Greenman and Xie's study points to the important role of social context: both processes and consequences of assimilation should depend on the local social context in which immigrants are embedded. They find little empirical evidence supporting hypotheses derived from segmented assimilation theory.

Brown (2007) studied social and spatial assimilation among several generations of Mexicans in the US based on a survey. She found that progress concerning both social and spatial assimilation was slow among the first and second generation, but significantly faster among the third generation. She pointed to 'family obligations' among the first generations as an important reason for their slow progress. Some had to send remittances back to family in Mexico or to house family coming to the US as illegal immigrants, which made their situation more difficult. She proposed a special model for Mexicans called *delayed assimilation*, where the essence is that some immigrants have to go through a longer process of deliberation from their inherited obligations to their ethnic origin before they can make progress in their new country.

In Europe the concept of integration of immigrants has had a somewhat different meaning in different countries (Phillips 2009; Schneider and Crul 2010). Immigration has in most countries had a much shorter history than in the US. In some countries in Northern Europe guest workers from Turkey and Pakistan first appeared in the 1960s and the first really big waves of refugees arrived around 1990. The first generation of immigrants is still heavily dominating.

Some immigrants did come to Europe because they wanted work and to earn money but not primarily because they wanted to settle permanently. This has especially applied to the immigrants who came as foreign labour from countries like Turkey and Pakistan back in the 1960s and 1970s. Moreover, some of the refugees, who were forced to leave their country in a hurry, expected to go back to where they came from. Such immigrants may not be very motivated for engaging themselves in a new country, besides getting jobs and earning money, and are more inclined to keep their orientation towards their country of origin. It could be expected that such immigrants could behave differently and have other residential preferences than immigrants who came to settle permanently. Diaspora could have a separate importance regarding location and choice of neighbourhood. It could reinforce immigrants' inclinations to separate from the native population and increase their wishes to live together in enclaves with countrymen or people with the same cultural background or religion. It could also affect their choice of housing in such a way that they would to a greater extent prefer to rent instead of buying a home. A high degree of diaspora might be assumed to result in a stronger wish to maintain the cultural patterns of their country of origin.

In a German study (Schaeffer and Bukenya 2010), it was concluded that language is one of the most critical factors for determining integration and assimilation at the workplace and in society. Ersanilli and Koopmans (2010) found in a study comparing Germany, France and the Netherlands that host culture adoption is best measured by host country identification, host country language proficiency and use, and interethnic social contacts. In a Danish public white paper (Tænketanken 2007), the factors which were found most important for integration were education and language skills, labour market participation, being financially self-supporting, absence of discrimination and social contacts between Danes and immigrants in daily life.

In contrast to the US, Europe consists of many different countries with large differences in integration contexts, which include differences in institutional arrangements in education, the labour market, housing, religion and legislation. These differences have a significant impact on the kind and speed of immigrants' integration (Crul and Schneider 2010). It is to be expected that there is a connection between general welfare policies and efforts towards integration of immigrants. Countries with highly developed welfare policies might also use more resources on helping immigrants to adapt to their new country.

It has also been debated whether the national, often negative, discourses in countries on the quality of immigrants and the effects of immigration in general influence integration. They play an important role in who is considered integrated or assimilated in both the public and the academic debate. The implicitly formulated 'ideals of integration or assimilation' differ greatly across countries (Crul and Schneider 2010). Some authors (Schaeffer and Bukenya 2010) are of the opinion that negative attitudes by natives towards immigrants have affected integration. Others (Crul and Schneider 2010) think that practices at the level of daily interactions sometimes even openly contradict official stances and rhetoric. A lot of pragmatism can be observed in the ways in which state agencies and societal institutions in Europe respond to the specific needs of immigrants and their children. They propose a *comparative integration context theory*, which argues that participation in social organisations and belonging to local communities across European cities are strongly dependent on the integration context organised by local authorities and NGOs.

Some European studies support the classical assimilation theory, while others do not. A Danish study (Skifter Andersen 2006) showed a strong connection between measures of integration (language proficiency, participation in the labour market and social contacts to Danes) and duration of stay in the country. Kalter and Granato (2001) concluded in a German study that 'in spite of the fact that the educational gap (between immigrants and natives) has clearly widened over the years under observation, it will turn out that in respect of the other aspects of life the general trend appears to be towards assimilation, especially for the second generation of the "classical" labour migrants'. Luthra (2009) concluded from his study that second-generation guest workers in Germany, in particular

Turks, Iberians and Greeks, show a significant integration advantage compared to the first generation. But there are considerable differences between different ethnic groups. Safi (2008) conducted a French study which could not support the existence of a uniform convergence process for different immigrant groups and concluded that 'Other, more complex, segmented models seem to characterize the various communities represented in the survey'.

Luthra (2009) discussed the opposite hypothesis of the assimilation theory; that some immigrant groups tend to develop a subculture or to end up in permanent deprivation which tends to isolate them from society. He concluded: 'Thus, the predictions of divergence – of compounded disadvantage (or, poor outcomes for poorly received groups) as well as compounded advantage (accelerated progress for positively received groups) does not appear to hold in the German case. I therefore conclude that ethnic origins do matter, but not in the ways consistent with assimilation theories as they are applied in the United States'.

As has been shown in this section, perceptions of immigrants' cultural, social and economic integration, and evidence about it, differ greatly. The dominant view is that assimilation theory still has some validity for most immigrants, but that integration, and the speed of it, differs greatly across countries and ethnic groups. There is evidence that some immigrants do not enter into mainstream society, but integrate into an ethnic underclass, what has been called downward assimilation.

Changes in immigrants' preferences for neighbourhood/tenure with duration of stay

The third assumption upon which the spatial assimilation theory is based is that immigrants change their preferences for housing and neighbourhood in pace with their integration over time.

In the qualitative study among British Asians by Phillips et al. (2007), the interviewees often acknowledged that especially younger people, often descendants, would gradually move into more ethnically mixed neighbourhoods in the suburbs. Members of multigenerational households as well as younger families had been motivated to move for better housing, schools and neighbourhoods, and to fulfil a desire for more privacy and independence from their ethnic group. For some, the move was an important means for establishing a social as well as a physical distance from less educated families in the inner-city areas. Others articulated their reasons for moving in terms of improvements in life-style and life-chances and increased opportunities for mixing with families from other ethnic groups.

Chapter 3 referenced a Danish study, which showed a strong connection between immigrants' preferences for living in multi-ethnic neighbourhoods or ethnic enclaves and their social and economic integration measured by their language proficiency, their participation in the labour market and their social contacts with Danes. It was also concluded that these neighbourhood preferences influenced the actual settlement pattern. This was also shown in a European study with data from 13 countries (Glikman and Semyonov 2012). If immigrants' integration improves over time, it should thus be expected that their preferences for neighbourhood also will change. Especially it is to be expected that improvements of the economic situation lead to stronger preferences for better housing in more prosperous neighbourhoods, usually in homeownership.

As in the US, spatial assimilation in Europe often is connected with moving into homeownership. Homeownership is a sign of economic success. As described in Chapter 2, many studies also point to a link between homeownership and social and cultural integration or assimilation (Alba and Logan 1992; Borjas 2002), as it indicates long-term economic progress and plays a key role in providing financial security (Sinning 2010) and it may also indicate immigrants' degree of commitment to the host country (Constant et al. 2009). The German study of Constant et al. (2009) showed a connection between homeownership and measures of integration. It was based on survey data on knowledge of the German language, preferred media, social relations and migration plans, which were used as indicators of integration. These indicators were shown to affect significantly the probability of homeownership, showing assimilated and integrated immigrants to be much more likely to own their place of residence. It was demonstrated that those with a stronger commitment to and identification with the host country were more able to move into homeownership for a given set of socio-economic and demographic characteristics and housing market conditions.

While there are several studies on the connection between the actual location of ethnic minorities and their integration over time, which will be discussed in the next section, studies of changes in immigrants' stated preferences over time, or between generations, are rare. Van der Laan Bouma-Doff (2007) tested in a Dutch study, with a statistical model, if preferences for living among co-ethnics differed between immigrants who had been less than 10 years in the country and those who had stayed more than 10 years. She did not find a statistically significant difference between the two groups except for among one single ethnic group. She did not, however, test for interaction with other variables in the model that might have been correlated with integration and preferences for neighbourhood, like language proficiency and social contacts with natives. The influence of these variables could have suppressed the effect of the length of stay. She also found that acculturation variables like immigrant status, command of the Dutch language and contact with native Dutch seemed to have little impact on preferences for living close to co-ethnics.

Studies of spatial assimilation

In North America, there has been intensive research about the settlement and segregation of immigrants, where the spatial assimilation theory has been central. The spatial assimilation model formulated by Massey (1985) has generally been supported empirically in a large number of American studies published before 1990 (see an overview in Alba and Logan 1992). Later studies have questioned parts or all of the theory. Especially the earlier assumption that new immigrants settle in ethnic enclaves in central city slums and later move to better housing in the suburbs has been questioned. Others have pointed to major differences across immigrants belonging to different ethnic groups concerning spatial assimilation. Some have found the concept of segmented assimilation applicable to the spatial settlement of immigrants over time.

It has been found that some newer immigrants settle directly in neighbourhoods in the suburbs (Clark and Patel 2004; Logan et al. 1999) and that some of them move into ethnic enclaves located there. Myles and Hou (2003) found e.g. that especially Chinese immigrants in Canada quickly purchased homes in more affluent neighbourhoods with somewhat more affluent and enduring ethnic communities, but also that other groups followed the traditional pattern, starting with settlement in the central cities followed by moves to suburbs. Others have also found that for some immigrant groups the central city-suburb model is still valid. Alba et al. (1999) found support for some aspects of the theory. The determinants of suburban residence were partly consistent with the spatial assimilation model.

Others have questioned the idea of ethnic enclaves as the way new immigrants are settling in the US. Instead of settling into a single sociocultural community, immigrants settle in a mosaic of self-sustaining ethnic communities containing several ethnic groups (called the pluralist model). Zelinsky and Lee (1998) have questioned both this model and the general spatial assimilation model. They argue that the assumptions concerning that immigrants only can make ethnic communities by settling in ethnic enclaves does not fit with reality. They formulated a theory that an ethnic network in the modern world quite easily can be established across neighbourhoods in a city, called heterolocalism.

Brown (2007) referenced several studies of Hispanic residential settlement patterns that tended to show results that appear consistent with classic spatial assimilation theory, but directed attention to group differences in the pace at which various aspects of spatial assimilation may be fulfilled. She also pointed out that the literature shows that divergence in the rapidity of spatial integration may occur within groups. Brown (2007) herself studied social and spatial assimilation among several generations of Mexicans in the US based on a survey. She found, in accordance with her findings about social assimilation, that progress concerning spatial assimilation was slow among the first and second generations, but significantly faster among the third generation. She concluded that substantial spatial integration did not emerge until the third generation. Several other studies have been done on spatial assimilation of different groups of Hispanics. Iceland and Nelson (2008) also concluded that Hispanics experience multiple and concurrent forms of spatial assimilation across generations. They showed that racial differences across different groups of Hispanics had importance for the pattern of spatial assimilation, in that Black Hispanics to a greater extent continued to be concentrated spatially. They found that race influences segregation despite the general strength of assimilation-related factors supporting the segmented assimilation perspective. Also, South et al. (2005) found variations in the residential mobility process among Latinos that are consistent with the segmented assimilation perspective. The generational and socio-economic differences that are anticipated by the classical assimilation model emerged more strongly for Mexicans than for Puerto Ricans or Cubans. Among Puerto Ricans and Cubans, darker skin colour resulted in lower mobility into white neighbourhoods (South et al. 2005).

European studies that throw light on processes of spatial assimilation are relatively sparse, and the literature in English is mostly from the Netherlands, Britain and the Scandinavian countries. As institutional conditions and ethnic composition of immigrants differ greatly between European countries, it should be expected that immigrants' integration and settlement patterns may differ greatly. Spatial assimilation might be very different across countries and ethnic groups.

Moreover, most European countries, except for former colonial powers, have had a relatively short history of immigration from Non-Western countries. In most countries the first generation of immigrants is heavily dominating and adult descendants are relatively few.

Only two studies have been found that use longitudinal data following immigrants' location over time from the date of arrival. A Dutch study (Zorlu and Mulder 2008) examines immigrants' settlement at the year of arrival and 5 years later. A Danish study, described at the end of this chapter, has used data on how immigrants settle when they arrive, and every year after, for up to 23 years. Glikman and Semyanov (2012) used survey data from 13 European countries to explore the connection between immigrants' length of stay and ethnic composition of their neighbourhood. They also examined the statistical connection between immigrants' income and ethnic composition of neighbourhood.

Other studies have looked at the change in neighbourhood for all, or a subset, of immigrants during a certain period and tested if these changes confirm spatial assimilation. Two studies did not include data on ethnic composition but measured the change in other neighbourhood variables during a period of time. Lersch (2013) examined the change in neighbourhood quality for families with Turkish background using data from a survey. Wessel et al. (2017) looked at the change from 2000 to 2008 in the social status of neighbourhoods for Non-Western immigrants in four Nordic countries using register data, making a comparative approach. Especially they tested if an improvement of social status led to an improvement of neighbourhood status.

Others have looked at how ethnic minorities relocate between neighbourhoods with different proportion of ethnic minorities. Schaake et al. (2014) assessed the impact of ethnicity, education and income among moving families on the change in socio-economic status and ethnic composition of the neighbourhood of residence, for three groups of ethnic minorities (Caribbean, Turks and Moroccans) who relocated in the Netherlands during one year. Two other studies, among Iranian and Iraqi immigrants, examined who moved out of multi-ethnic neighbourhoods in Stockholm (Macpherson and Strömgren 2013) and the Netherlands (Bolt and Van Kempen 2010). Moreover, a qualitative study in Britain (Phillips et al. 2007) has investigated the connection between immigrants' choice of location and their social integration. Finally some studies have looked at how the settlement of descendants differs between first generation of immigrants in Denmark (Skovgaard Nielsen 2016), the Netherlands (Zorlu and Mulder 2011; Bolt and

Van Kempen 2002) and in 13 European countries (Glikman and Semyanov 2012), testing the relevance of the theory of straight-line assimilation.

Bolt and Van Kempen (2002) used a questionnaire among 75 Turks and Moroccans in three neighbourhoods in the city of Utrecht to examine their housing and neighbourhood career. Among the residents in an old neighbourhood close to the city, they found that many had remained there for a long time and that most movers had re-housed within their neighbourhood. Residents in a newer multi-ethnic neighbourhood in the suburb were more dissatisfied with it and found that although there were many co-ethnics and some ethnic facilities, the neighbourhood did not have the same attraction as the old central neighbourhood. They had ended up in the new neighbourhood because the waiting lists for social rented dwellings were relatively short. They only intended to stay for a limited period in their housing career. Some families, who were at a later stage of their career, had moved to more ethnically mixed neighbourhoods. The main conclusion of the study was, however, that only a smaller portion of the two immigrant groups had moved to other kinds of neighbourhoods.

It is a well-known fact that ethnic minorities – and natives – living in multi-ethnic neighbourhoods are poorer than average. Glikman and Semyanov (2012) studied the connection between income and living in multi-ethnic neighbourhoods, with a majority of ethnic minorities, based on a survey with 3,825 respondents in the following 13 European countries: Belgium, Austria, Germany, Denmark, Sweden, Norway, Luxembourg, France, United Kingdom, Netherlands, Spain, Greece and Switzerland. They found, in all but one country, that ethnic minorities living in such neighbourhoods had lower incomes. The likelihood of residing in ethnic neighbourhoods tended especially to be higher among immigrants of non-European origin, and among immigrants belonging to the Muslim faith, even after taking into consideration the economic status of the immigrants. They also revealed that residence in ethnic neighbourhoods tends to decrease with the passage of time in the host country. They concluded that there is 'Firm support for the argument that immigrants tend to translate their economic resources into residential location'. The study showed, however, that different ethnic and cultural subgroups experience divergent patterns of residential attainment, even after taking into consideration the differences among groups in length of residence in the host country and in socio-economic characteristics. A quite surprising conclusion from the study was that the patterns of residential location observed were quite uniform across the countries, despite substantial differences among European countries in social, political and economic structure, welfare policy, housing policy and immigrant integration policy.

Some studies have looked at how ethnic minorities relocate in connection with residential moves. Lersch (2013) examined if Turks in Germany obtained improved neighbourhood quality (like less crime, noise and air pollution) when they moved across neighbourhoods and found that they most often moved between neighbourhoods with similar low quality on average, and that they were also significantly less likely than Germans to improve their neighbourhood through moving. He also found that income played a surprisingly small role in determining neighbourhood quality changes. Wessel et al. compared the housing situations of ethnic minorities in four capitals in Nordic countries in two different years with a gap of 8 years. It was found that the economic integration, measured by upward earnings mobility, in all countries was weak during the 8 years in question. They also concluded that movement from neighbourhoods with low economic status to neighbourhoods with higher status on average was lacking. Also was found a weak relationship between upward earnings mobility during the years and upward spatial mobility. Change in the ethnic composition of neighbourhoods was not included in the study, but the study does investigate if ethnic segregation in the cities influenced mobility in the period, which it did not. The two studies thus did not precisely relate to the spatial assimilation theory, which emphasises the ethnic character of neighbourhoods, but illuminated changes in characteristics which are assumed to accompany spatial assimilation. Both studies found a weak connection between income and neighbourhood improvements in a shorter time period, which could insinuate that many ethnic minority families within a shorter timespan tend to make their housing career inside certain neighbourhoods.

These results have, however, been contradicted by other studies. Schaake et al. (2014) assessed the impact of ethnicity, education and income on the change in socio-economic status and ethnic composition of the neighbourhood of residence for ethnic minorities who had relocated in one year in the Netherlands. The total effects of income and education for neighbourhood choice were positive for all ethnic groups, implying spatial assimilation. However, members of Non-Western minority ethnic groups showed a smaller increase in both socio-economic status and share of the native majority in the neighbourhood of residence than the members of the ethnic majority, also after controlling for education and income.

Bolt and Van Kempen (2010) looked at moves within one year from multi-ethnic neighbourhoods with more than 40 per cent Non-Western ethnics to neighbourhoods with less than that in the four largest cities of the Netherlands. They showed that income had a positive influence on moving from multi-ethnic neighbourhoods to other neighbourhoods and that education had an even stronger effect. Duration of stay in the Netherlands for the first generation did not, however, increase spatial assimilation. The weakness of the study is that it only contains two groups of neighbourhoods, while some of the moves could be between neighbourhoods with only small differences in the proportion of ethnic minorities.

This weakness was less pronounced in the study conducted by Zorlu and Mulder (2008), which followed the settlement of immigrants 5 years after their arrival to the Netherlands. They divided the neighbourhoods in the study into three groups: ‘white neighbourhoods’ with less than 10 per cent Non-Western ethnic minorities, ‘mixed neighbourhoods’ with 10 to 35 per cent Non-Western ethnic minorities and multi-ethnic neighbourhoods with more than 35 per cent Non-Western ethnic minorities. The results did not fit with the spatial assimilation theory, as there was a clear pattern of concentration into more segregated neighbourhoods among immigrants from Non-Western countries. Those who settled initially in a less segregated neighbourhood tended to move towards more segregated neighbourhoods in the four consecutive years. This is, however, a very short period after immigration, where social integration hardly has been completed. Moreover, it might be assumed that it is so difficult for new immigrants to get housing that they settle anywhere it is possible. The longitudinal Danish study described in the next section also shows that immigrants in the first years tend to settle more and more in multi-ethnic neighbourhoods, but that after a couple of years some of them leave again. The estimated differences in the residential mobility patterns of different immigrant groups, in the study conducted by Zorlu and Mulder (2008), confirmed, however, some predictions of spatial assimilation theory. Immigrants in paid employment moved more frequently out of multi-ethnic neighbourhoods and their destination was more likely to be a less segregated neighbourhood. Unemployed immigrants, in contrast, were likely to move into more segregated neighbourhoods. It was also shown that the settlement and mobility patterns of asylum migrants were markedly different from those of other migrants. They were initially much more spread over the country, probably as a result of specific government settlement policies, and their settlement pattern was not sensitive to local economic conditions. However, in time, asylum migrants showed the strongest tendency of all migrant groups to move to more ethnically segregated neighbourhoods. This particularly strong tendency implies that government settlement policies designed to regulate initial settlement locations have a limited impact.

Macpherson and Strömgren (2013) used data on two ethnic groups in Stockholm, of Iranians and Iraqis, for 2001 and 2007 to test if they had left multi-ethnic neighbourhoods in the period. They found that income, education and length of stay in Sweden influenced decisions to move out of multi-ethnic neighbourhoods. However, a possible source of errors was that some could have moved from one multi-ethnic neighbourhood to another.

A number of studies have looked at differences in spatial assimilation between first and second generation of immigrants, what has been called straight-line assimilation. Glikman and Semyanov's (2012) study in 13 European countries compared the frequency of living in multi-ethnic neighbourhoods for the two generations and found that residence in an ethnic neighbourhood in most countries was more prevalent among first generation immigrants than among second generation immigrants. In six countries, however, the effect of generation was positive but not statistically significant (Denmark, France, Greece, Luxembourg, Spain, Netherlands and Norway). In one country, Belgium, the effect was negative and not significant. The impact of 'generation' remained statistically significant even after controlling for socio-demographic characteristics, ethnic origin and religious affiliation of the immigrants.

The differences between first and second generation of immigrants were also compared by Bolt and Van Kempen (2010) in their study of immigrants' residential moves in the four largest cities of the Netherlands. They found that the second generation moved more frequently into non-concentration neighbourhoods than the first generation. Skovgaard Nielsen (2016) studied the moving pattern of young Somali and Turkish adults leaving home in Denmark to examine to what extent there can be found a link between segregation of parents and segregation of their grown-up children and if the pattern of young immigrants differed from that of descendants. She uncovered that both young immigrants and descendants settled more seldom in multi-ethnic neighbourhood than their parents; the share of home-leavers settling in these neighbourhoods was lower than for their parents, also when controlling for background variables like income and education. The probability for settling in a multi-ethnic neighbourhood was, however, higher if they had grown up in such a neighbourhood, pointing to intergenerational effects. This effect did not disappear when the economic situation of parents was introduced in the statistical model. Thus, there was no indication that the parental economic situation impacts on the neighbourhood outcome of the home-leaver and thus no indication that a better economic situation of the parents leads to less likelihood of moving into an ethnic neighbourhood. Somewhat surprisingly, parental employment situation did impact on the home-leaving of their children, but a clear effect for moving to an ethnic or a non-ethnic neighbourhood was not found. Turkish immigrants were more affected than Turkish descendants by the share of ethnic minorities in the parental neighbourhood. This could indicate that home-leaving descendants get more spatially assimilated than home-leaving immigrants. Moreover, major differences between the residential patterns of the two ethnic groups were found. Young Somalis leaving home were much more likely to move to multi-ethnic neighbourhoods than young Turks.

Evidence from Danish studies of spatial assimilation

Some of the assumptions on spatial assimilation put forward by the previously referenced literature have been examined in two Danish studies. The first was based on a survey among recently moved Non-Western ethnic minorities in 2005. It looked at why some of them moved into multi-ethnic neighbourhoods and at the differences between those who moved in and those who left the neighbourhoods. In- and out-movers to and from multi-ethnic neighbourhoods were examined (Skifter Andersen 2010).

The other study was based on a database that contained information on ethnic minorities and a sample of Danes, and their housing, every year from 1985 to 2008. The data contained information on age, income, length of stay, housing, neighbourhood composition, etc. for every person every year in the period. Denmark was divided into 9,000 neighbourhoods with an average of about 600 inhabitants. Data on every neighbourhood was constructed; in particular the proportion of residents belonging to different ethnic minorities (immigrants and their descendants). For more information, see Skifter Andersen 2015 and 2016. This data made it possible to illuminate immigrants' situation on arrival in Denmark and the development over time in their residential situation in up to 25 years. Moreover, it was possible to examine the change that occurred in the neighbourhoods over time. Both studies made it possible to test if the spatial assimilation theory is valid in Denmark, a Social Democratic welfare state.

Denmark is a relatively new immigration country. The number of immigrants and descendants from Non-Western countries increased from 150,000 in 1980 to 490,000 in 2009. The proportion of immigrants in Denmark rose to about 10 per cent in 2010. A little less than half were immigrants from non-European countries and about 1 per cent were from Eastern and Central Europe. Denmark has a large social housing sector, which it is relatively easy to get access to, even if some waiting time occurs in large cities. In general it may be said that immigrants in Denmark, compared with many other countries (Skifter Andersen et al. 2013; Skovgaard Nielsen et al. 2015), have had quite easy access to social housing. A large part of them (54 per cent) have made their housing career in social housing, where they have been able to find housing that satisfied their housing needs.

Changes over time in immigrants' family situation and integration

A large proportion of immigrants coming to Denmark since 1985 did not have their own dwelling when they first settled in the country, as can be seen from Figure 4.1.

About 50 per cent of newly arrived Non-Western immigrants had to live in extended families with other people than their own nuclear family just after they had obtained a residence permit, some in refugee camps, others together with different kinds of family or friends. There was no difference between immigrants coming from different parts of the world. Within some years after, most of them moved out and obtained homes for themselves. After 6 years, still 10 per cent were left in extended families.

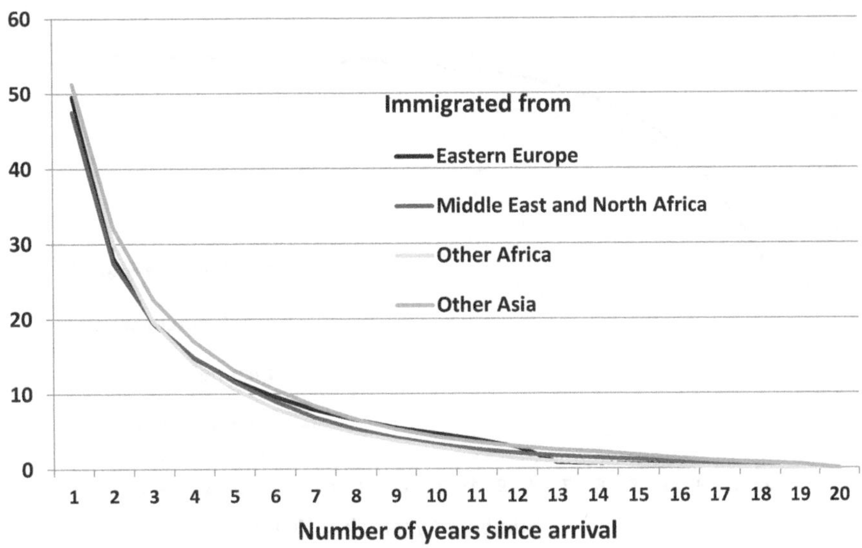

Figure 4.1 The proportion of Non-Western immigrants living in extended families, etc.; sharing their dwelling with other than the nuclear family, depending on the number of years they have been in Denmark, 1985–2007

Source: Skifter Andersen 2015

The Danish study shows an improvement in the economic situation of Non-Western immigrants in Denmark during their first years in the country, but also that after some time the economic integration seems to stall and that Non-Western immigrants never reached the same level of prosperity as the majority population. The employment rate and income of immigrants were quite low in the first years, as can be seen from Figure 4.2, which shows the development in immigrants' employment rate and income compared to the figures for comparable Danes of the same age and sex.

Newly arrived men had on average an employment rate that was only about 30 per cent of the employment for Danes. During the first 15 years it increased to 50 per cent and stagnated thereafter. Employment of women was somewhat lower, starting with 20 per cent and stagnating at 40 per cent. In the same period, the relative income of men rose from 40 per cent of comparable Danes' to more than 60 per cent. Somewhat surprisingly, the income of immigrant women became closer to that of Danish women, rising from 35 per cent to 70 per cent.

The settlement of newly arrived immigrants

In accordance with the spatial assimilation theory, we should expect that most newly arrived immigrants to Denmark would settle in multi-ethnic neighbourhoods in social housing. The settlement pattern of new Danish immigrants does

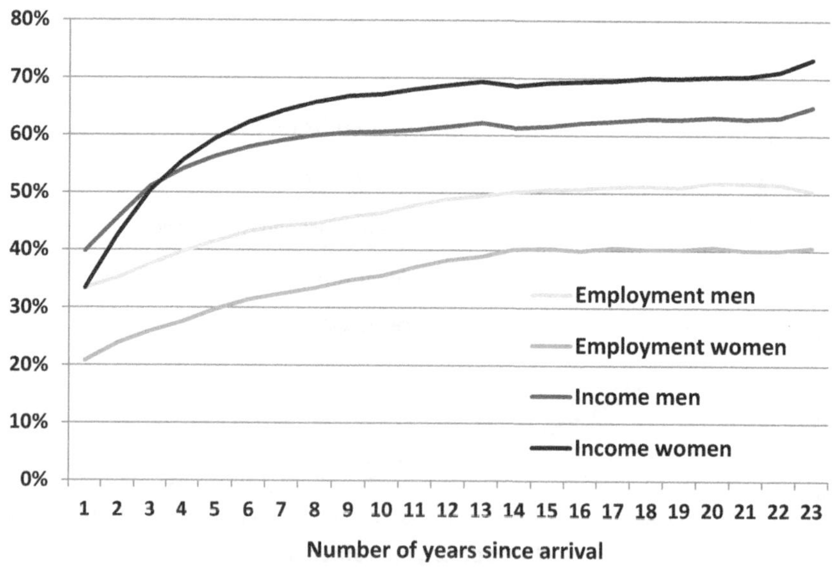

Figure 4.2 The development in Non-Western immigrants' economic integration during their first 23 years in Denmark, 1985–2008. The relative employment rate and income of immigrants as a proportion of that of Danes with the same age and sex

Source: Skifter Andersen 2015

not, however, quite respond to the pattern expected from the theory. Thirty-eight per cent of the Non-Western newcomers in the period 1985 to 2007 settled in social housing. This is less than all Non-Western minorities, with 54 per cent. Only 18 per cent settled directly in multi-ethnic neighbourhoods, with more than 20 per cent ethnic minorities, while 40 per cent of all minorities live in such neighbourhoods.

There is, however, great variation in the pattern. Different immigrants are in very different positions when they arrive. Besides their education and ability to get work, their family situation and how they came to the country are important for their housing options and choice of neighbourhood. Especially immigrants who were granted family reunification found themselves in another situation because they moved in with their family, who had settled earlier in the country and had already taken the first steps in a housing career. Figure 4.3 shows the proportion of immigrants arrived within the actual year that settled directly into social housing.

This data shows that the assumption in the spatial assimilation theory that new immigrants start their housing career in ethnic neighbourhoods does not quite fit the situation in a country like Denmark. The data indicates that immigrants who came as family unified with other ethnic minorities to a much higher degree settled in social housing and in multi-ethnic neighbourhoods. Among the other immigrants, more than 40 per cent of families with children settled in social housing, but only about 12 per cent in multi-ethnic neighbourhoods. This can probably

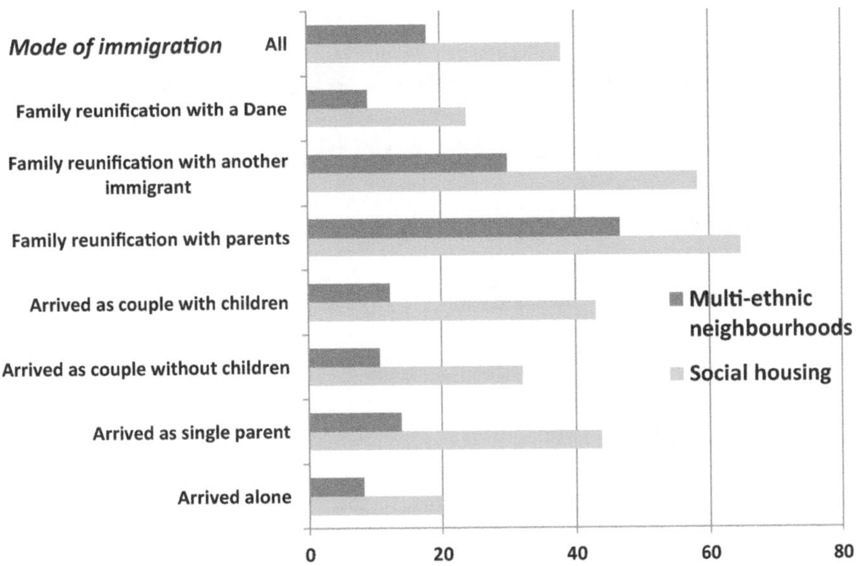

Figure 4.3 Proportion (%) of newly arrived Non-Western immigrants, 1985–2008, who settled in social housing or multi-ethnic neighbourhoods with more than 20 per cent Non-Western ethnic minorities, depending on their mode of immigration

Source: Skifter Andersen 2015

partly be ascribed to the asylum system in Denmark, where new refugees are located by local authorities. Among immigrants arriving as singles, only 20 per cent moved into social housing and 8 per cent to multi-ethnic neighbourhoods.

A statistical model was created to reveal the factors determining if newly arrived immigrants settled in multi-ethnic neighbourhoods (for details see Skifter Andersen 2015). It contained data on mode of arrival and for family reunification, the number of years the family had been in the country, country group of origin, age, sex, family situation, size of family, employment situation, household income, urbanisation and time period. The four country groups used were East Europeans, West Asia/North Africa, Other Asia and Other Africa. Figure 4.4 shows the importance of the mode of arrival for the probability of moving directly into multi-ethnic neighbourhoods with more than 20 per cent Non-Western ethnic minorities just after arrival.

The figure depicts the results of the model used to calculate the increased probability of living in multi-ethnic neighbourhoods compared to immigrants who have been family reunified with Danes. All the other immigrants more often have moved into multi-ethnic neighbourhoods from the start, especially those who have been family reunified; and the longer their reunified family has been in the country, the higher probability for moving to multi-ethnic neighbourhoods. Other

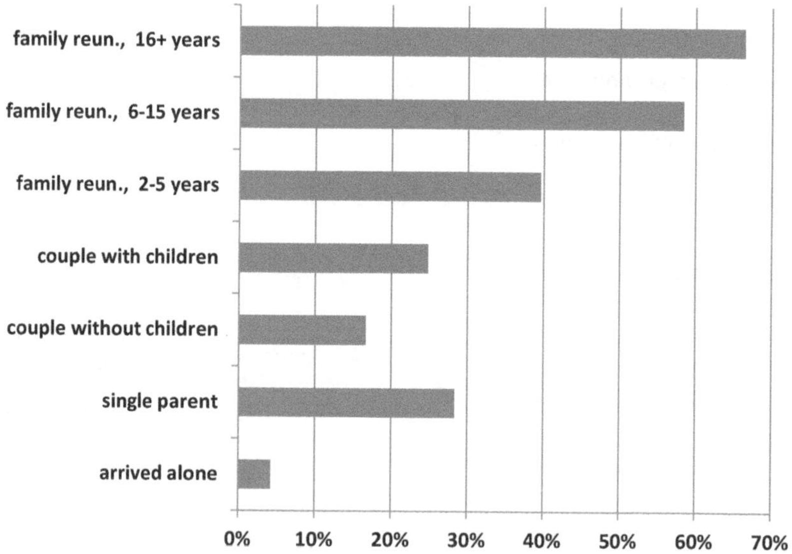

Figure 4.4 Increased probability that newly arrived Non-Western immigrants in Denmark, 1985–2008, moved into a multi-ethnic neighbourhood with more than 20 per cent Non-Western minorities, compared with immigrants who have been family reunified with Danes, depending on mode of arrival and duration of stay for the earlier immigrants receiving families by reunification

Source: Skifter Andersen 2015

immigrants have much more seldom moved directly into multi-ethnic neighbourhoods, especially those who arrived as a single individual without family. There were also some differences between ethnic groups, as immigrants from the Middle East and Africa much more often settled in multi-ethnic neighbourhoods from the beginning.

Comparison of ethnic minorities moving to and from multi-ethnic neighbourhoods

The theory of spatial assimilation assumes that new immigrants often move into enclaves with many countrymen but that some of them only stay there for a couple of years and move out when they have become more integrated in the new society.

To partially test this hypothesis, we have compared households moving to and from multi-ethnic neighbourhoods with more than 40 per cent of residents belonging to Non-Western ethnic minorities. A simple comparison confirms the hypothesis because it shows that newcomers less often have work (see Table 4.1) than households moving out.

Table 4.2 shows the results from a statistical analysis (logistic regression) comparing newcomers and households moving out of the areas. The model tries to determine the variables explaining if a moving household is a newcomer to multi-ethnic neighbourhoods or is moving out. The sample consists only of ethnic minority households moving to and from multi-ethnic neighbourhoods.

The constructed model is not very successful in explaining the variation in differences between newcomers and households moving away. This indicates that other important variables explaining why people move are lacking in the model, for example family and job changes. But there are some significant variables that tend to confirm the spatial assimilation theory. Newcomers have a higher probability of being refugees, they have more seldom obtained Danish citizenship, they tend to have lower incomes, are more often on welfare payments and more seldom in employment. They are younger and more often have children. Descendants of immigrants, on the other hand, tend less often to move into multi-ethnic neighbourhoods, which is in accordance with the results of Skovgaard Nielsen (2015 and 2016).

Table 4.1 Comparison of newcomers, households moving out, internal movers and residents among ethnic minorities in all multi-ethnic neighbourhoods with more than 40 per cent Non-Western ethnic minorities in Denmark, 2002

	In employment	*Without work*	*Others**	*Total*
	- % -			
Newcomers	31	56	13	100
Households moving out	48	37	15	100
Internal moves	30	58	11	100
Residents	40	46	13	100

* Students and pensioners.
Source: Skifter Andersen 2010

Table 4.2 Results from a logistic regression analysis of the difference between ethnic minority households moving to and from multi-ethnic neighbourhoods with more than 40 per cent Non-Western ethnic minorities – increased probability for being a newcomer

	Increased probability (%)
Has Danish citizenship?	−37
Descendant – e.g. born in Denmark?	−28
Refugee?	22
Age divided by 10	−10
Have children?	14
Household income/100.000 DKK	−14
On welfare?	34
In employment?	−20*
Nagelkerke R2	0.038

Source: Skifter Andersen 2010

All results are significant within the 5 per cent limit. * This result is only significant when income is omitted from the model.

Immigrants' settlement over time after arrival

As far back as the 1970s, immigrants in Denmark began to concentrate in certain parts of the cities. In the first years, they settled in the older parts with cheap deteriorated private rented housing in the centre of the cities. As the immigrant population grew and many of these neighbourhoods underwent urban renewal, immigrants settled more and more in neighbourhoods in suburbs with social housing. Partly due to the increasing number of immigrants and their descendants, the proportion of ethnic minorities increased greatly in some neighbourhoods. The number of neighbourhoods with more than 20 per cent ethnic minorities from Non-Western countries increased from 60 in 1985 to nearly 900 in 2008 (out of 9,000 neighbourhoods). As more and more neighbourhoods over time exceeded 20 per cent, it might be expected that the proportion of immigrants living in such neighbourhoods would be steadily increasing with the duration of their stay in Denmark.

Figure 4.5 shows the connection between the duration of stay in Denmark for Non-Western immigrants and the share of them living in social housing or in multi-ethnic neighbourhoods with more than 20 per cent Non-Western minorities. Only immigrants who did not come by family reunification, and who immigrated to Denmark after 1985, are included.

It can be seen that dwelling in social housing increased during the first 10 years of stay, peaking at about 67 per cent, and then it began to decrease until about 50 per cent for those who had lived in the country the longest time, 23 years. As Zorlu and Mulder (2008) found, the proportion living in multi-ethnic neighbourhoods increased during the first couple of years from a little more than 10 per cent to nearly 50 per cent after 15 years. Hereafter the presence declined to 36 per cent after 23 years. The study (Skifter Andersen 2016) determined that minorities who

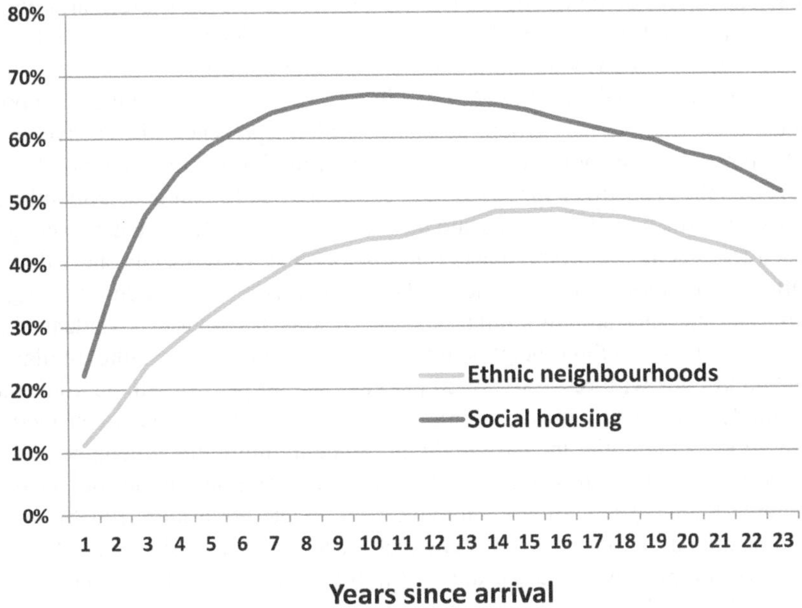

Figure 4.5 The proportion of Non-Western immigrants, who were not family reunified, living in social housing and multi-ethnic neighbourhoods with more than 20 per cent Non-Western minorities depending on duration of stay in Denmark, 1985 to 2008

Source: Skifter Andersen 2015

originated from Muslim and African countries were more likely to settle in multi-ethnic neighbourhoods, but all ethnic groups followed the same pattern over time of increasing presence in the first 10–15 years of residence in the country followed by stagnation and decline.

The increase in immigrants' presence in multi-ethnic neighbourhoods in Denmark during the first 10–15 years might be a result of two different developments. It might be because immigrants moved to multi-ethnic neighbourhoods during those years, or it might be because they stayed in a neighbourhood where the proportion of immigrants was increasing. To determine to what extent the development was a result of changing neighbourhoods or of actual choices, two statistical analyses were made, one for all immigrants (who were not family reunified) and one for those who had moved within the previous year. The analyses examined the connection between residing in multi-ethnic neighbourhoods (determined as more than 20 per cent Non-Western ethnic minorities) and duration of stay, divided into five groups. To control for differences between periods of time, and for other background factors, logistic regressions were made with the same control variables as those used previously for newly arrived immigrants (country of origin group, age, sex, family situation, size of family, employment situation, household income, urbanisation and time period). To test the importance of the somewhat arbitrary definition of multi-ethnic neighbourhoods, another model was constructed for living in neighbourhoods with more than 40 per cent minorities.

Figure 4.6 shows the increased probability of living in multi-ethnic neighbourhoods for duration of stay groups compared with the presence when arrived.

All three analyses point to some spatial assimilation within the first 23 years of stay, as the probability for living in or moving to multi-ethnic neighbourhoods increased in the first 10–15 years and thereafter declined. Because they were controlled for income (and other background variables), these results show that the spatial assimilation that took place in Denmark was not only a consequence of increasing incomes among immigrants. The results are quite independent of the definition of multi-ethnic neighbourhoods, as the tendency is quite similar for neighbourhoods with respectively 20 per cent and 40 per cent ethnic minorities. The model for recently moved immigrants shows, moreover, that a considerable proportion of the initial increase in the presence in multi-ethnic neighbourhoods has been a result of an increase in the proportion of ethnic minorities in some neighbourhoods. The results for moving immigrants are a more reliable measure of spatial assimilation. They show that the tendency to reside in multi-ethnic neighbourhoods only increased during the first 6–10 years of stay, thereafter it stagnated, and after 11–15 years it began to fall.

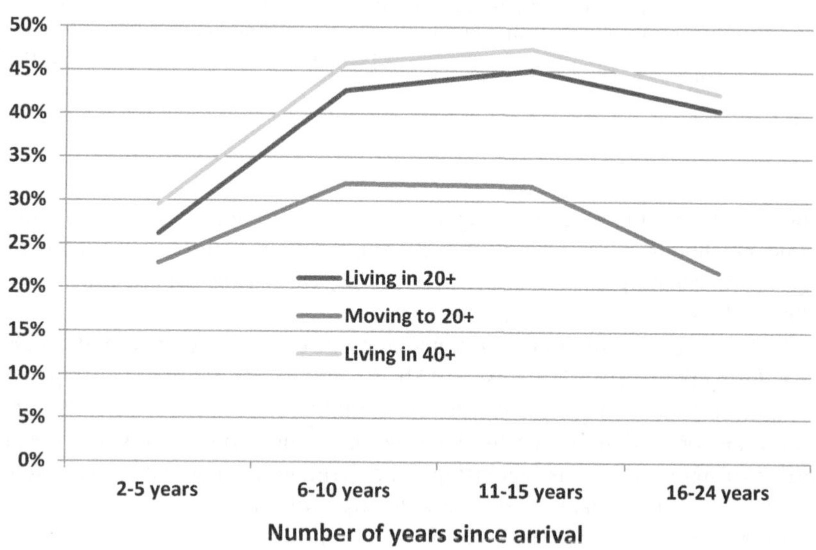

Figure 4.6 Increased probability of living in multi-ethnic neighbourhoods in duration of stay groups compared with the presence in the first year after arrival, 1985–2008, for living in neighbourhoods with more than 20 per cent and more than 40 per cent, respectively, Non-Western ethnic minorities, and for those who had moved to the neighbourhoods within the previous year

Source: Skifter Andersen 2016

References

Alba R. D. and Logan J. R. (1992) Assimilation and Stratification in the Homeownership Patterns of Racial and Ethnic Groups, *International Migration Review*, 26(4), pp. 1314–1341.

Alba R. D., Logan R. J. and Stults B. J. (2003) Residential Inequality and Segregation in an Immigration Era: An Analysis of Major U.S. Metropolitan Regions in 1990. In Reitz J. *Host Societies and the Reception of Immigrants*, pp. 119–148. La Jolla: Center for Comparative Immigration Studies.

Alba R. D., Logan J. R., Stults B. J., Marzan G. and Zhang W. (1999) Immigrant Groups in the Suburbs: A Reexamination of Suburbanization and Spatial Assimilation, *American Sociological Review*, 64(3), pp. 446–460.

Alba R. D. and Nee V. (1997) Rethinking Assimilation Theory for a New Era of Immigration, *International Migration Review*, 31(4), pp. 826–874.

Bolt G., Ozuekren A. S. and Phillips D. (2010) Linking Integration and Residential Segregation, *Journal of Ethnic and Migration Studies*, 36(2), pp. 169–186.

Bolt G. and Van Kempen R. (2010) Ethnic Segregation and Residential Mobility: Relocations of Minority Ethnic Groups in the Netherlands, *Journal of Ethnic and Migration Studies*, 36(2), pp. 333–354.

Bolt G. and Van Kempen R. (2002) Moving Up or Moving Down? Housing Careers of Turks and Moroccans in Utrecht, the Netherlands, *Housing Studies*, 17(3), pp. 401–422.

Borjas G. J. (2002) Homeownership in the Immigrant Population, *Journal of Urban Economics*, 52, pp. 448–476.

Brown S. K. (2007) Delayed Spatial Assimilation: Multigenerational Incorporation of the Mexican-Origin Population in Los Angeles, *City & Community*, 6(3), pp. 193–209.

Brubaker R. (2001) The Return of Assimilation? Changing Perspectives on Immigration and Its Sequels in France, Germany, and the United States, *Ethnic and Racial Studies*, 24(4), pp. 531–548.

Clark W. A. W. and Patel S. (2004) *Choices of the Newly Arrived Foreign Born: Spatial Patterns and the Implications for Assimilation.* On-Line Working Paper Series, University of California, CaliforniaCentre for Population Research.

Clifford J. (1999) Diasporas. In Vertovec S. and Cohen R. *Migration, Diasporas and Transnationalism.* Cornwall: Edward Elgar Publishing Limited.

Cohen R. (1997) *Global Diasporas*. London: UCL Press.

Constant A. F., Roberts R. and Zimmermann K. F. (2009) Ethnic Identity and Immigrant Homeownership, *Urban Studies*, 46(9), pp. 1879–1898.

Conzen K., Gerber D., Morawska E., Pozzetta G. and Vecoli R. (1992) The Invention of Ethnicity: A Perspective from the U.S.A., *Journal of American Ethnic History*, 12(3), pp. 3–41.

Crul M. and Schneider J. (2010) Comparative Integration Context Theory: Participation and Belonging in New Diverse European Cities, *Ethnic and Racial Studies*, 33(7), pp. 1249–1268.

Ersanilli E. and Koopmans R. (2010) Rewarding Integration? Citizenship Regulations and the Socio-Cultural Integration of Immigrants in the Netherlands, France and Germany, *Journal of Ethnic and Migration Studies*, 36(5), pp. 773–791.

Gilroy P. (1996) Diasporaen og identitetens omveje (Diaspora and the Detours of Identities), *Social Kritik* (Copenhagen) 45/46, pp. 19–41.

Glazer N. and Moynihan D. P. (1970) *Beyond the Melting Pot: The Negroes, Puerto Ricans, Jews, Italians, and Irish of New York City*. Cambridge, MA: MIT Press.

Glikman A. and Semyanov M. (2012) Ethnic Origin and Residential Attainment of Immigrants in European Countries, *City & Community*, 11(2), pp. 198–219.

Gordon M. (1964) *Assimilation in American Life*. New York: Oxford University Press.

Greenman E. and Xie Y. (2008) Is Assimilation Theory Dead? The Effect of Assimilation on Adolescent Well-being, *Social Science Research*, 37, pp. 109–137.

Haan M. (2007) The Homeownership Hierarchies of Canada and the United States: The Housing Patterns of White and Non-White Immigrants of the Past Thirty Years, *International Migration Review*, 41(2), pp. 433–465.

Iceland J. and Nelson K. A. (2008) Hispanic Segregation in Metropolitan America: Exploring the Multiple Forms of Spatial Assimilation, *American Sociological Review*, 73, pp. 741–765.

Kalter F. and Granato N. (2001) *Recent Trends of Assimilation in Germany*. Fakultät für sozialwissenschaften ZUMA. Mannhein: Universitaet Mannheim.

Lee J. and Bean F. D. (2004) America's Changing Colour Lines: Immigration, Race/Ethnicity, and Multiracial Identification, *Annual Review of Sociology*, 30, pp. 221–242.

Lersch P. M. (2013) Place Stratification or Spatial Assimilation? Neighbourhood Quality Changes after Residential Mobility for Migrants in Germany, *Urban Studies*, 50, pp. 1011–1029.

Logan J. R., Zhang W. W. and Alba R. D. (2002) Immigrant Enclaves and Ethnic Communities in New York and Los Angeles, *American Sociological Review*, 67(2), pp. 299–322.

Luthra R. R. (2009) *Assimilation in a New Context: Second Generation Educational Attainment in Germany*. On-Line Working Paper Series, University of California, California Center for Population Research.

Macpherson R. A. and Strömgren M. (2013) Spatial Assimilation and Native Partnership: Evidence of Iranian and Iraqi Immigrant Mobility from Segregated Areas in Stockholm, Sweden, *Population Space and Place*, 19, pp. 311–328.

Massey D. S. (1985) Ethnic Residential Segregation: A Theoretical Synthesis and Empirical Review, *Sociology and Social Research*, 69, pp. 315–350.

Model S. (1991) Caribbean Immigrants: A Black Success Story? *International Migration Review*, 25, pp. 248–276.

Myles J. and Hou F. (2003) *Neighbourhood Attainment and Residential Segregation among Toronto's Visible Minorities. Analytical Studies Branch Research Paper Series*. Business and Labour Market Analysis Division Statistics Canada.

Nagel C. R. (2009) Rethinking Geographies of Assimilation, *The Professional Geographer*, 61(3), pp. 400–407.

Park R. E. (1928) Human Migration and the Marginal Man, *American Journal of Sociology*, 33, pp. 881–893.

Park R. E. and Burgess E. (1969) *Introduction to the Science of Sociology*, 1921 Reprint. Chicago, IL: University of Chicago Press.

Phillips D. (2009) Minority Ethnic Segregation, Integration and Citizenship: A European Perspective, *Journal of Ethnic and Migration Studies*, 36(2), pp. 209–225.

Phillips D. (2007) Moving Towards Integration: The Housing of Asylum Seekers and Refugees in Britain, *Housing Studies*, 21(4), pp. 539–554.

Phillips D., Davis C. and Ratcliffe P. (2007) British Asian Narratives of Urban Space, *Transactions of the Institute of British Geographers*, 32(2), pp. 217–234.

Portes A. and Rumbaut R. G. (Eds.) (2001) *Legacies: The Story of the Immigrant Second Generation*. Berkeley, CA: University of California Press.

Portes A. and Zhou M. (1993) The New Second Generation: Segmented Assimilation and Its Variants, *Annals of the American Academy of Political and Social Science*, 530, pp. 74–97.

Robinson D., Reeve K. and Casey R. (2007) *The Housing Pathways of New Immigrants*. Joseph Rowntree Foundation.

Robinson D. (2008) *New Immigrants and Migrants in Social Housing in England Local Government Association*. Centre for Regional Economic and Social Research, Sheffield: Sheffield Hallam University.

Safi M. (2008) The Immigrant Integration Process in France: Inequalities and Segmentation, *Revue française de sociologie*, 49(5), pp. 3–44.

Schaake K., Burgers J. and Mulder C. H. (2014) Ethnicity, Education and Income, and Residential Mobility between Neighbourhoods, *Journal of Ethnic and Migration Studies*, 40(4), pp. 512–527.

Schaeffer P. V. and Bukenya J. O. (2010) Assimilation of Foreigners in Former West Germany, *International Migration Early View*, 52(4), pp. 157–174.

Schneider J. and Crul M. (2010) New Insights into Assimilation and Integration Theory: Introduction to the Special Issue, *Ethnic and Racial Studies*, 33(7), pp. 1143–1148.

Scholten P., Baggerman F., Dellouche L., Kampen V., Wolf J. and Ypma R. (2017) *Policy Innovation in Refugee Integration*. Rotterdam: Ministerie van Sociale Zaken en Werkgelegenheid.

Sinning M. (2010) Homeownership and Economic Performance of Immigrants in Germany, *Urban Studies*, 47, pp. 387–409.

Skifter Andersen H. (2016) Spatial Assimilation? The Development in Immigrants' Residential Career with Duration of Stay in Denmark, *Journal of Housing and the Built Environment*, 31(2), pp. 297–320.

Skifter Andersen H. (2015) *Indvandring, integration og etnisk segregation. Udviklingen i indvandrernes bosætning siden 1985*. Copenhagen: Danish Building Research Institute.

Skifter Andersen H. (2010) Spatial Assimilation in Denmark. Why Do Immigrants Move to and from Multi-ethnic Neighbourhoods? *Housing Studies*, 25(3), pp. 281–300.

Skifter Andersen H. (2006) *Bo sammen eller spredt? Etniske minoriteters boligønsker og motiver for bosætning (Living together or apart. Motives for choice of residence among ethnic minorities)*. Horsholm: Danish Building Research Institute.

Skifter Andersen H., Magnusson Turner L. and Søholt S. (2013) The Special Importance of Housing Policy for the Housing Situation of Ethnic Minorities. Evidence from a Comparison of Four Nordic Countries, *International Journal of Housing Policy*, 13(1), pp. 20–44.

Skovgaard Nielsen R. (2016) Straight-line Assimilation in Leaving Home? A Comparison of Turks, Somalis and Danes, *Housing Studies*, 31(6), pp. 631–650.

Skovgaard Nielsen R. (2015) Leaving Home in Denmark: A Comparison of Differences between Turks, Somalis and Danes, *Journal of Housing and the Built Environ*, 30, pp. 613–630.

Skovgaard Nielsen R., Holmquist E., Dhalmann H. and Søholst S. (2015) Comparing Somalis' Perceived Possibilities in the Nordic Capital Housing Markets: Local Context and Cultural Background, *Housing Studies*, 30(3), pp. 433–452.

Slack T. and Jensen L. (2007) Underemployment Across Immigrant Generations, *Social Science Research*, 36, pp. 1415–1430.

South S. J., Crowder K. and Chavez E. (2005) Migration and Spatial Assimilation among U.S. Latinos: Classical Versus Segmented Trajectories, *Demography*, 42(3), pp. 497–521.

Texeira C. and Drolet J. (2016) *Settlement and Housing Experiences of Recent Immigrants in Small- and Mid-Sized Cities in the Interior of British Columbia, A Pathways to Prosperity Project*. Vancouver, University of British Columbia.

Tænketanken (2007) *Værdier og normer blandt udlændinge og danskere (values and norms among foreigners and Danes)*. Copenhagen: Ministeriet for flygtninge, indvandrere og integration.

Valdez Z. (2006) Segmented Assimilation among Mexicans in the Southwest, *Sociological Quarterly*, 47(3), pp. 397–424.

Van Der Laan Bouma-Doff W. (2007) Involuntary Isolation: Ethnic Preferences and Residential Segregation, *Journal of Urban Affairs*, 29(3), pp. 289–309.

Waldinger R. and Feliciano C. (2004) Will the New Second Generation Experience 'Downward Assimilation? Segmented Assimilation Reassessed, *Ethnic and Racial Studies*, 27(3), pp. 376–402.

Waters M. C. (1999) *Black Identities: West Indian Immigrant Dreams and American Realities*. Cambridge, MA: Harvard University Press.

Wessel T., Andersson R., Kauppinen T. and Skifter Andersen H. (2017) Spatial Integration of Immigrants in Nordic Cities: The Relevance of Spatial Assimilation Theory in a Welfare State Context, *Urban Affairs Review*, 53(5), pp. 812–842.

Wright R., Ellis M. and Parks V. (2005) Re-placing Whiteness in Spatial Assimilation Research, *City and Community*, 5(2), pp. 111–135.

Zelinsky W. and Lee B. A. (1998) Heterolocalism: An Alternative Model of the Sociospatial Behaviour of Immigrant Ethnic Communities, *International Journal of Population Geography*, 4, pp. 281–298.

Zhou M. (1997) Segmented Assimilation: Issues, Controversies, and Recent Research on the New Second Generation, *The International Migration Review*, 31(4), pp. 975–1008.

Zorlu A. and Mulder C. H. (2011) Ethnic Differences in Leaving Home: Timing and Pathways, *Demography*, 48, pp. 49–72.

Zorlu A. and Mulder C. H. (2008) Initial and Subsequent Location Choices of Immigrants to the Netherlands, *Regional Studies*, 422, pp. 245–264.

5 Selective migration of ethnic minorities and natives in connection with ethnic neighbourhoods

Ethnic segregation is not only a result of the behaviour of ethnic minorities but also of the behaviour of the native born population. The question is; which kind of moving behaviour has the dominant importance for the creation and maintenance of ethnic segregation and multi-ethnic neighbourhoods? Is it the behaviour of ethnic minorities or the behaviour of natives? Is it because ethnic minorities more often move into multi-ethnic neighbourhoods or because they tend to remain there when they have got access? Or is it because natives move away, when ethnic minorities move in, or because they do without choosing these neighbourhoods?

In this chapter is examined how the migration of both groups is influenced by the ethnic composition of neighbourhoods or other factors. The possible causes of selective migration among ethnic minorities have been illuminated in the previous chapters. Here we will focus more in detail on reasons for selective migration among natives in European countries, in the literature called 'white flight and avoidance'.

The moving behaviour of natives has traditionally been shown to be influenced by their preferences for living in ethnically mixed or homogenous neighbourhoods. It could be assumed that such preferences are affected by to what extent natives are exposed to ethnic minorities depending on the size of the population of ethnic minorities and their spatial concentration in cities. The so-called contextual contact literature points towards that living close to other ethnic groups diminishes resentment towards such groups. At the same time, the perceptions of ethnic minorities among people who do not live in multi-ethnic neighbourhoods are negatively affected by the size of the ethnic population in general.

Explanations for preferences among natives for avoiding neighbourhoods dominated by other ethnic groups have constantly been discussed in the literature. In the American literature so-called racial prejudices have been a central subject. Some have proposed that it is not only resentment towards other groups that is important, but that these groups are connected to a lower social status. It has also been proposed that it is not so much the ethnic composition of neighbourhoods that is significant, but more the social character of the neighbourhood. Some researchers have proposed that the decisive reason for selective migration among natives is the social deprivation which often is found in ethnic neighbourhoods, what has been called *the racial proxy theory*. Also, the character of local schools might have importance.

Evidence from different studies on selective migration, and its possible causes, will be described, especially the results from a survey among natives in three Nordic countries. This study shows widespread reluctance among natives against living in neighbourhoods with many immigrants, but the attitudes differ greatly across the countries. One of the important reasons is the conditions in local schools. Moreover, results from a Danish quantitative study on actual moves in a period of 25 years to and from neighbourhoods with different proportions of ethnic minorities will be referenced in the last part of the chapter. The study examines if it is the behaviour of ethnic minorities or the behaviour of natives that has been the most important cause of the appearance of multi-ethnic neighbourhoods and ethnic segregation in Denmark. Both Nordic studies indicate that the behaviour of natives is the most important. The social composition of neighbourhoods has some importance, but is not decisive.

Selective migration among the native majority and its causes

Some of the European research literature on ethnic spatial segregation and concentration has focused on the importance of what has been called ethnic selective migration, that is, ethnic differences between those leaving, entering and staying in so-called multi-ethnic neighbourhoods (Andersson 2013; Andersson and Bramå 2006; Magnusson and Özüekren 2002; Bolt et al. 2008; Bolt and Van Kempen 2010). It is assumed that moving behaviour among both ethnic minorities and natives (in the US, white Europeans) in relation to multi-ethnic neighbourhoods differs from the behaviour displayed in other neighbourhoods, and that this divergent behaviour creates and maintains spatial concentrations of ethnic minorities and ethnic segregation in the cities.

In this section we will especially discuss reasons for selective migration among natives. Divergent behaviour among natives has been denoted as *white flight* if natives tend to have a higher mobility out of multi-ethnic neighbourhoods, and *white avoidance* if they less often move into multi-ethnic neighbourhoods than to similar neighbourhoods without ethnic minorities. The concepts originally stem from the US, where they denote a tendency for white Americans to leave or avoid neighbourhoods with an increasing proportion of African-Americans or other ethnic minorities (Quillian 2002; Wright et al. 2005; Crowder et al. 2011; Crowder and South 2008; Pais et al. 2009; Farley et al. 1997; Clark 2002).

Evidence of white flight and avoidance has also been found in some European countries. Andersson (2013) showed in a study in Stockholm that Swedes tend to avoid moving to multi-ethnic neighbourhoods, but he did not find evidence of white flight. He showed clear evidence that native-born Swedes are less inclined than most immigrant categories to move into immigrant-dense areas, while ethnic origin did not seem to matter much when explaining who leaves such areas. Andersson and Bramå (2006) found some evidence of white flight in Stockholm, but also concluded that white avoidance was by far the most important process. Vilkama (2011) showed that despite Helsinki's much lower proportion of immigrants, both flight and avoidance are significant also in this city. Bolt et al. (2008), in a study of migration in the Netherlands, concluded that 'Dutch and Western immigrants are much more likely than non-Western groups to move out of concentration neighbourhoods and to opt more often for neighbourhoods with a low proportion of minorities' and this 'can only to a very limited extent be attributed to differences in socioeconomic status'. Another Dutch study (Schaake et al. 2010) similarly concluded that 'Native Dutch move away more frequently than ethnic minorities as the share of non-western ethnic minorities in neighbourhoods is greater'. In the Danish study described in the last section, based on an extensive database on moves during 25 years, clear evidence of both white flight and avoidance is found.

As discussed in more detail in Chapter 2, studies of reasons for migration and choice of neighbourhood can be based on either 'stated' or 'revealed' preferences. Revealed preferences are deducted from the actual choices of neighbourhood, which are found by examining how people actually live and move around. Stated preferences are determined by directly asking people where they prefer to live and why. Both methods involve methodological problems. In research based on revealed preferences, the possible reasons for them are deducted by statistical analyses of how neighbourhood choice is connected to available data on households and neighbourhoods, for example the proportion of ethnic minorities in the neighbourhood. Most of the studies on white flight and avoidance have been based on such data. The main problem with this research is that it is difficult to explain why people choose where to live and if the ethnic composition of the neighbourhoods has importance.

The reasons for white flight and avoidance have been examined in several studies in the US by help of surveys and qualitative research. In Europe the most extensive study is made by Semyonov and Glikman (2009) and Semyonov et al. (2007), who used survey data from 20 countries obtained from the 2002 European Social Survey. Andersson et al. (2017) made a survey among the native majority in the three Nordic capitals: Oslo, Stockholm and Helsinki. Kaufmann and Harris (2015) used data on political voting and moving behaviour to test whether people voting for the Conservative party, who were assumed to be more critical against ethnic minorities, more often showed evidence of white flight and avoidance. Several studies in Europe (Permentier et al. 2011; Skifter Andersen 2006; Van Der Laan Bouma-Doff 2007; Havekes et al. 2014), which have looked at reasons for dissatisfaction with neighbourhoods, have included questions about if this is influenced by the ethnic composition of residents. Feijten and van Ham (2009) studied how white flight is influenced by changes in neighbourhood conditions. They examined how a resident's expression of a wish to move from the neighbourhood was influenced by changing conditions in the neighbourhood concerning socio-economic status, proportion of ethnic minorities and population turnover.

Reasons for white avoidance might be very different from flight. White avoidance is based on people's imaginations of how it is to live in an ethnic enclave or a multi-ethnic neighbourhood and on a publicly dispersed reputation of these neighbourhoods, often communicated by the media in a distorted way. In the case of flight, the movers have themselves experienced how it is to live in such a neighbourhood and have a specific knowledge about the living conditions in them. While there is an ample assortment of research on why residents want to leave their neighbourhood, European research on why people want to avoid specific neighbourhoods is relatively sparse. Most studies on white avoidance thus examine revealed preferences based on quantitative data on moves between neighbourhoods where one tries to control for all other variables than the ethnic composition of neighbourhoods. This is often a difficult task. The results could be distorted because some important variables that are correlated with ethnic composition are missing in the analyses.

Preferences for ethnic homogeneity or ethnic mix in neighbourhoods

In the US, comprehensive research has shown that some white Americans especially prefer to avoid or leave neighbourhoods dominated by African-Americans, but also neighbourhoods with some other ethnic groups (Ellen 2000; Quillian 2002; Wright et al. 2005; Crowder et al. 2011; Crowder and South 2008; Pais et al. 2009; Farley et al. 1997; Clark 2002).

In Europe comparative research on this is somewhat scarce. Semyonov et al. (2007) used survey data from the 2002 European Social survey (ESS) from 20 European countries on preferences for living in ethnic homogeneous neighbourhoods. Moreover, the respondents were asked if they currently lived in homogeneous neighbourhoods. In the analyses made by Semyonov et al. (2007), dummy variables were introduced for countries that, despite the control for individual variables, revealed considerable cross-country variation in preferences for residing in ethnically homogeneous areas. It was also shown that there is no straightforward connection between such preferences and the proportion of natives actually living in such neighbourhoods. It is difficult to explain this by geographical variables or by differences in the character of the welfare states (see Chapter 8).

Semyonov et al. (2007) also conducted statistical analyses on personal factors determining preferences, which gave support to the hypothesis that a preference for homogeneous residence tends to be more pronounced among socio-economically weak and vulnerable populations and among individuals with conservative political views. Respondents with better education and income less often had high preferences, while unemployment status increased them. Also, older people had higher preferences.

Are preferences influenced by the general level of immigration and ethnic segregation in countries?

It could be assumed that preferences for living in ethnically homogenous neighbourhoods are influenced by to what extent natives are exposed to ethnic minorities, depending on the size of the population of ethnic minorities and their spatial concentration in cities. At least two studies are able to throw light on how differences across European countries, in the extent of ethnic minorities and the degree of ethnic segregation, affect preferences for living in ethnically homogenous neighbourhoods.

Andersson et al. (2017) conducted a comparative study in three Nordic capitals among the native majority in Stockholm, Oslo and Helsinki with surveys that asked for the opinions among natives on whether an ethnic mix in neighbourhoods is desirable or not. It should be expected that cities with higher segregation also would have weaker preferences for ethnics mixing among natives. The comparison of the Nordic cities points somewhat unexpectedly towards a reverse connection between level of ethnic segregation and preferences for ethnic homogeneity. Stockholm has the strongest ethnic segregation while it is lowest in Helsinki. The study showed thus that support for ethnic mixing in neighbourhoods was strongest in the most segregated city, Stockholm (67 per cent), and weakest in the least segregated city, Helsinki (43 per cent).

The same conclusion appeared from the study conducted by Semyonov et al. (2007) (see Chapter 8). It showed that even if the picture is not quite clear, the least segregated countries, measured as the smallest proportion not living in homogenous neighbourhoods, also tend to have the strongest preferences for ethnic homogeneity.

Another possible explanation for high preferences for homogeneous neighbourhoods and low for ethnic mixes could be to what extent countries have a high proportion of ethnic minorities. Natives who are more exposed to ethnic minorities might have higher preferences for homogeneity. In the Nordic study by Andersson et al. (2017), the proportion of ethnic minorities also varied greatly across the three cities and the ethnic composition differed. Especially Helsinki has a much smaller ethnic population which is dominated by immigrants from Eastern Europe. Preferences for ethnic mixing were, however, highest in Stockholm with the largest population of ethnic minorities and lowest in Helsinki with the smallest.

The importance of the size of the immigrant population can also be illuminated by the data from Semyonov et al. (2007) by comparing the proportion of foreign born residents in countries with the measured preferences for ethnic homogeneous neighbourhoods. This comparison (see Chapter 8) shows that there is not a clear connection between preferences for ethnically homogeneous neighbourhoods and the extent of immigration to the countries. It seems, however, that countries with high immigration less often tend to prefer ethnic homogeneity, and countries with fewer ethnic minorities more often tend to prefer it. This is in accordance with the referenced Nordic study.

The effect of personal contacts with other ethnic groups

Natives in countries with a high proportion of ethnic minorities are more often exposed to ethnic minorities in their neighbourhood. The important question is to what extent this has an influence on their preferences for the ethnic composition of their neighbourhood. The Nordic study examined both the attitudes to ethnically mixed neighbourhoods among residents living in multi-ethnic neighbourhoods and among people living outside. Moreover, they examined the differences between people who had moved recently from these neighbourhoods and people who had not. In general preferences for ethnics mixed among natives living in multi-ethnic neighbourhoods were not very different from those living outside multi-ethnic neighbourhoods. In Oslo there was a small tendency to higher support of ethnics mixing among natives living in multi-ethnic neighbourhoods. This is in accordance with the study conducted by Semyonov et al. (2007). They showed that preferences for ethnic homogeneity were much lower among those who currently lived in neighbourhoods where some or most residents were members of ethnic minorities, also when controlling for personal background factors like income and employment.

The possible consequences of living in multi-ethnic neighbourhoods have been studied in the so-called contextual contact literature. This extensive research has been concerned with the question whether mixing of different ethnic groups in neighbourhoods promotes more tolerance between the groups or causes conflicts

(Wessel 2009). As referenced in Andersson et al. (2017), American research shows that the implications of personal contacts between people are different from the impacts of more impersonal contacts found in larger urban areas than neighbourhoods, so-called contextual contacts. Research about personal contacts between the majority population and minorities shows that more personal contacts tend to reduce negative attitudes against minorities among the majority (Pettigrew and Tropp 2008). Research about contextual contact as found in larger urban areas, however, points towards increased negative attitudes. There seems to be greater resentment against ethnic minorities in cities with a larger population of ethnic minorities. Stein et al. (2000) conclude thus: 'Ironically, our findings suggest that the very conditions [increasing minority density] that give rise to white hostility toward minority group members set in motion a corrective for this hostility: intergroup contact'.

Kaufmann and Harris (2015) found a similar pattern by a review of 24 studies. The researchers themselves found, in a study in England and Wales, that opposition to immigration among natives was lower in ethnically diverse neighbourhoods (wards), but in larger geographical areas (municipalities) there was a reverse effect. This evidence indicates that a general dislike for minorities has less importance for white flight than for white avoidance, which is more influenced by the composition of residents in larger areas of cities. Geographical proximity thus promotes intergroup contact, but it could also promote racial hostility.

A possible reason for higher tolerance for ethnic minorities in mixed neighbourhoods could be that this is a result of selective migration, where the least tolerant residents have left. Kaufmann and Harris (2015) examined the hypothesis, but could only find a weak confirmation for it.

Explanations for aversions towards ethnically mixed neighbourhoods

Extensive research in the US has shown that white Americans mostly prefer to settle in neighbourhoods with few black people or other ethnic minorities (e.g. Clark 2002; Pettigrew and Tropp 2006; Bobo and Zubrinsky 1996; Farley et al. 1997; Ellen 2000), despite the large variation in residential preferences within the group – from those who are relatively tolerant to those who feel threatened by a few black neighbours (Farley et al. 1997; Galster 1990; Swaroop and Kryzan 2011).

The role of prejudices towards ethnic minorities

Several assumptions have been formulated about the reasons why natives in Europe and North America (in the US white Americans) have aversions towards living in ethnically mixed neighbourhoods. An important question is if natives avoid or leave multi-ethnic neighbourhoods because they have a general antipathy against some minority ethnic groups and are averse to living amongst them. Explanations for preferences among white Americans for avoiding neighbourhoods dominated by other ethnic groups have constantly been discussed. In the American literature so-called racial prejudices have been a central subject (Pettigrew and Tropp 2006; Bobo and Zubrinsky 1996; Farley et al. 1997; Clark 2002). The core of this explanation is that preferences for location are influenced by prejudices about other ethnic groups. Bobo and Zubrinski found strong support for this hypothesis among whites. Also Farley et al. (1994) argue that racial prejudice and conflict underlie residential preferences of whites. In the Netherlands the connection between interethnic prejudice and residential preferences was studied by Van Der Laan Bouma-Doff (2007). She found that preferences for co-ethnic neighbours were driven mostly by interethnic prejudice and that this to a certain extent contributed to observed ethnic residential isolation.

Even if some of the results from the US have been confirmed in studies in Europe, it must be assumed that reasons for white flight and avoidance differ in Europe. It must also be presumed that flight and avoidance behaviour differ across European countries, among others because the extent of anti-minority attitudes varies with the extent and composition of migration, the historic development and the political conditions in countries. Good data on differences in attitudes towards ethnic minorities across European countries has been provided through the survey data from the 2002 European Social survey (ESS) from 20 countries. Semyonov and Glikman (2009) used these data to illuminate differences across countries concerning anti-minority attitudes in the populations. They calculated what they called a '*perception of threat index*,' which was constructed from a couple of survey questions on fear of negative consequences of immigration like competition for jobs, public expenses, cultural concerns and crime. The data shows pronounced differences across the countries in the extent of negative attitudes towards ethnic minorities. As shown in Chapter 8, countries with a high perception of threat also tend to have stronger preferences for ethnic homogeneity in neighbourhoods. To the extent that these preferences lead to white avoidance, it can thus be concluded that ethnic segregation is influenced by the specific political and economic conditions in countries.

Clark (2002) states that the two structural attributes most often used as determinants of negative attitudes towards minority populations are the relative size of the ethnic group population and the economic conditions that prevail in the host society. Specifically, the larger the size of the minority population and the more depressed the economic conditions are, the greater the feeling of threat and fear of competition over scarce social resources and economic rewards. Competitive threat, in turn, is likely to produce greater hostility and prejudice towards minority populations. Semyonov and Glikman (2009) in their study conducted statistical analyses, based on survey data from the ESS 2002, that provided firm support for the idea that negative attitudes towards ethnic minorities are more pronounced in countries with a larger number of non-Europeans than in countries with a small number of non-Europeans. They also showed that the economic situation, measured by GDP, exerts a negative impact on perception of threat, lending support to the theoretical expectation that economic prosperity is likely to decrease negative attitudes towards ethnic minorities.

It is possible, however, that preference for homogenous neighbourhoods is not as much a result of racial prejudices as of a desire for cultural homogeneity, also called ethnocentrism. Semyonov et al. (2007) found that individuals in European countries who consider national and cultural homogeneity to be important also prefer to live in neighbourhoods with their 'own kind' – that is, with those who share their national and cultural identity. Thus, ethnocentrism and prejudice coexist and interplay in Europe in shaping individuals' preference to live apart from ethnic minorities.

Some have proposed that it is not only resentment towards other ethnic groups as such that is important, but that some groups are connected to a lower social status (Blumer 1958); that people have developed a perception of different ethnic groups as having different social status. It is the magnitude of social differences that in-group members have socially learned to expect and maintain relative to members of specific out-groups that influences their preferences for where to live, especially among the white majority, but also among ethnic minorities. Bobo and Zubrinsky (1996) examined what they called the *perceived social class difference hypothesis*. It says that negative preferences for living in neighbourhoods dominated by other ethnic minorities are caused by a perception of these neighbourhoods as inhabited by low-status groups. Bobo and Zubrinski did not, however, find strong evidence that perceived differences in status in general played a role in residential integration attitudes.

The racial proxy hypothesis. The importance of social deprivation and low social status in ethnic neighbourhoods

It has been proposed that it is not so much the ethnic and cultural content of neighbourhoods that is significant, but more the social character of the neighbourhood. The decisive reason for selective migration among natives is then the perceived social deprivation, which often is found in ethnic neighbourhoods.

Multi-ethnic neighbourhoods are often deprived housing estates with many residents outside the labour market and with visible social problems, and sometimes antisocial behaviour and crime. Several studies have shown that such problems influence residents' satisfaction with the neighbourhood and increase mobility. The assumption that this is the main reason for white flight and avoidance has been called the *racial proxy hypothesis* (Ellen 2000; Harris 2001; Dekker 2013; Swaroop and Kryzan 2011).

In the literature on mobility, general dissatisfaction with dwelling or neighbourhood has been seen as an important reason for the mobility process, but there are many different understandings of this concept and of what elements are contained in housing and neighbourhood satisfaction or dissatisfaction. As stated by Burrows and Rhodes (1999), dissatisfaction will always be a function of both objective realities and sometimes highly subjective evaluations. Research on causes of neighbourhood dissatisfaction shows that personal perceptions of housing and neighbourhood rather than 'objective' neighbourhood characteristics are keys for understanding whether people are satisfied or not, and whether they tend to stay put or move away (Lu 1998; Permentier et al. 2011).

Residential satisfaction is often split up into housing and neighbourhood satisfaction. There is evidence, however, that the two are not independent. The studies of Parkes et al. (2002) and Dekker et al. (2007) thus showed that there was a connection between the two kinds of satisfaction – people dissatisfied with their house being much more likely to be dissatisfied with their neighbourhood. One explanation could be that people who have decided to move – and have the possibility to do so – in general have created a negative evaluation of both dwelling and neighbourhood. Housing dissatisfaction and plans to move could therefore contribute to explaining neighbourhood dissatisfaction.

Many studies have found that fear of crime or lack of safety are very important for neighbourhood satisfaction, where high incidents of crime can considerably reduce the usually high neighbourhood satisfaction among residents. There is not always a direct connection between actual crime rates and the perceptions of local crime among residents (Basolo and Strong 2002). The perceived crime is found to be more important to satisfaction than the actual crime. As stated by Ærø (2006) and Costa Pinho (2000), the feeling of safety is not only connected to observed crime but also to conflicts between residents or groups of residents, grounded in experiences with other residents not acting in accordance to what is felt as usual norms and behaviour. Such experiences can lead to a general antipathy against other ethnic groups. This is an argument supporting that a socially and ethnically heterogeneous neighbourhood has a higher probability of neighbourhood dissatisfaction.

Most of the referenced studies on neighbourhood satisfaction have been conducted in typical neighbourhoods with relatively few ethnic minorities, where the importance of the ethnic composition of the neighbourhood has not been examined. The few studies that have been made in multi-ethnic neighbourhoods, based on surveys among natives, have not resulted in quite consistent conclusions. Harris (2001) concluded from his study of some American neighbourhoods with different proportions of African-Americans that there was an 'overwhelming support for the racial proxy hypothesis'. Preferences for neighbourhoods that were relatively affluent, safe, well-maintained and had good schools were decisive for higher satisfaction, and there was not a separate effect of the ethnic composition of the neighbourhood. Dekker (2013) also found that dissatisfaction with the neighbourhood among residents in multi-ethnic neighbourhoods in the Netherlands was not strongly connected to the ethnic composition. Other studies, however, have found that neighbourhood satisfaction is strongly related to ethnicity. Swaroop and Kryzan (2011) conclude from their study in Chicago that 'The racial proxy perspective applies weakly to whites' satisfaction: whites report less satisfaction in neighbourhoods with more minority residents, and only some of their dissatisfaction can be attributed to local social characteristics'.

A possible motive for white flight could be dissatisfaction with the social cohesion and social capital in a multi-ethnic neighbourhood. Andersson et al. (2017) refer to the discussion of Robert Putnam's (2007) study claiming that ethnic diversity in communities and neighbourhoods undermines trust between people. Putnam's study and conclusions have been disputed, but some American studies have confirmed that the trust of 'white Americans' in their neighbours is negatively influenced by ethnic diversity in the neighbourhood (Abascal and Baldassari 2015).

The importance of image and reputation

The meaning of neighbourhoods as places is socially constructed, and different groups give these places different meanings. A concept that has been used in some studies of residential satisfaction is the 'image' of the neighbourhood. A possible definition of image is 'the set of meanings by which an object is known and through which people describe, remember and relate to it. This is a net result of the interaction of a person's beliefs, ideas, feelings and impressions about an object' (Rijpers and Smeets 1998). Thus the image is a simplified generalised picture that summarises all the good and bad conditions in a neighbourhood. It is also to some extent assumed by the individual that other people have the same or a very similar image of the neighbourhood.

Rijpers and Smeets (1998) identified three types of images: 1. *The internal image*, which was the one that was common among residents of the neighbourhood, 2. *The external image*, according to people living outside the neighbourhood, and 3. *The self-reflecting image*, which was the image residents believed was found among outsiders. Their study found some differences between what conditions in the neighbourhood were important in these three different kinds of images. The internal image was mostly influenced by the social composition of residents, the visual quality of the built environment, social inconveniences and accessibility of facilities. The external image was mostly explained by social inconveniences (like vandalism, safety and behaviour of residents) and by quality of buildings. Finally, the self-reflecting image was somewhat similar to the internal image, but social inconveniences and visual qualities were more important. This study found that the external image did not coincide with the residents' perception of the neighbourhood – the so-called self-reflecting image. Unfortunately, the ethnic composition of the neighbourhood was not included in the study.

Another concept used in the literature is *reputation*. Reputation in many ways is a similar concept to image. A possible definition is 'The beliefs or opinions that are generally held about someone or something' (Permentier et al. 2007). This underlines the collective character of reputation. It is a widely shared belief, shared by certain groups. Reputation does not depend on objective neighbourhood attributes, but rather on the composite 'brand' or 'name' of the area. Reputation is often seen as an 'outside view of the area' (Koopman 2007). Permentier et al. (2007) also argue that the concept of reputation most often has been used when studying the attitudes and behaviour of non-residents.

Neighbourhood reputation can be used as an indicator of the status of residents. Poor reputation and low status connected with deprived neighbourhoods can thus be perceived as an important problem for the residents and a motive for moving to another neighbourhood. In this context it is residents' belief in what is the reputation among outsiders that has importance. This corresponds to what has been called the self-reflecting image. It is not directly determined by residents' perception of good and bad things in the neighbourhood.

The ethnic composition of a neighbourhood has been shown to have a marked influence on its reputation. Permentier et al. (2007) found that ethnic composition of neighbourhoods as well as average neighbourhood income were the strongest determinants of perceived reputation. Also an earlier Danish study based on data from 500 social housing estates with social problems (Skifter Andersen 2002) showed that the most important factors explaining poor reputation were proportion of ethnic minority groups together with visible social problems like social conflicts, noisy residents and problems with vandalism.

The reputation of a neighbourhood can influence plans to move and actual moves. Some earlier studies (Tsfati and Cohen 2003 – referenced in Permentier et al. 2007) indicate that perceived reputation has a separate importance for moving behaviour. A study in deprived neighbourhoods in Denmark by Skifter Andersen (2008) showed that the perception of the neighbourhood as having a poor reputation among outsiders was the most important reason for plans to move away. It can thus be assumed that perceived low personal status by living in deprived neighbourhoods is a very important reason for leaving the areas, and that the presence of a high concentration of ethnic minorities leads to a negative image and status. This can influence actual moving behaviour among residents in multi-ethnic neighbourhoods, but it can be assumed that the effect is even greater among people living outside.

The results from the previously referenced study in three Nordic capitals (Andersson et al. 2017) also indicate that residents in multi-ethnic neighbourhoods are very conscious about the negative impact on the image of a neighbourhood connected to a concentration of ethnic minorities. When inquiring about satisfaction with the current ethnic composition of the respondents' own neighbourhood, it appeared that residents in multi-ethnic neighbourhoods much more often than residents in other neighbourhoods wanted to have a lower share of ethnic minorities in their neighbourhood despite a more positive attitude to ethnic mixing in neighbourhoods. This can be explained by the fear of a negative image for the neighbourhood. The study also showed that those who had moved away did not have the same wish for a reduction in the share of minorities.

It has, however, also been questioned if dissatisfaction is an important reason for leaving a neighbourhood. Several studies have found that dissatisfaction not always results in plans to move and actual moves (Lu 1998; Clark and Ledwith 2005; de Groot et al. 2007; Kearns and Parkes 2003). In general it is known from the research literature on reasons for residential moves that dissatisfaction with the neighbourhood is seldom the main reason for a move, which mainly is triggered by housing demand and work changes (Skifter Andersen 2008). There are diverging conclusions on this. Lu (1998) showed a strong connection between moving intentions and neighbourhood satisfaction. Clark and Ledwith (2005), however, found in a study among households in Los Angeles that in general the effects of neighbourhood variables, beyond the structural effects of age and housing needs on moving intentions, were quite small. Kearns and Parkes (2003) showed that moving intentions mostly were linked to dissatisfaction with the dwelling, but also that perceived 'disorder' and dissatisfaction with people in the neighbourhood were good predictors of plans to move.

Boschman et al. (2017) examined the reasons why residents who were dissatisfied and wanted to leave their neighbourhood realised their desires or not. For neither Dutch natives nor ethnic minorities did the ethnic composition of the neighbourhood impact their desire to leave the neighbourhood.

Some researchers argue that it is not neighbourhood characteristics per se that influence residential mobility, but neighbourhood change (Lee et al. 1994) or expectations about future change. Feijten and Van Ham (2009) studied in the Netherlands whether neighbourhood change – increasing presence of Non-Western immigrants – affected the probability that residents wished to leave their neighbourhood. They found that ethnic change played a role, but the relationship turned statistically insignificant after controlling for residents' subjective opinions about such change. Natives staying in immigrant-dense neighbourhoods tended to have individual characteristics that are often found to co-vary with less positive views on immigrants, that is, they had on average less education and lower income, which corresponds to the findings by Semyonov et al. (2007). The study by Kearns and Parkes (2003) and van Ham and Clark (2009) showed that neighbourhood decline – the perception of things getting worse – had great importance for moves. This was also shown in another Dutch study (Feijten and van Ham 2009) in different neighbourhoods where there was a higher propensity to move among residents who perceived a decline in their neighbourhood.

The importance of local schools

An often-mentioned reason for moving away from multi-ethnic neighbourhoods, both among natives and minorities, is that schools in such neighbourhoods often have a lower quality, which motivates middle income families and especially native families with children to move away from or avoid these neighbourhoods.

A growing number of studies have tried to study the complex interplay of schools and neighbourhoods, bringing the school issue into research on neighbourhood preferences, residential mobility and white flight. Most of the studies are carried out in the US context (e.g. Clark and Maas 2012; Cascio 2012; Zhang 2011; Ledwith and Clark 2007), but there are also studies from the Netherlands (Gramberg 1998; Boterman 2012), the UK (Burgess et al. 2005; Johnston et al. 2006) and the Nordic countries (Rangvid 2009; Brattbakk et al. 2013). They all find some kind of connection between school problems and neighbourhood dissatisfaction or wishes to move away.

Rangvid (2009) found that Danes are more likely to opt out of their local public school if it has a large concentration of immigrant pupils. Danes are far more likely to leave as soon as the concentration exceeds 35 per cent. Among ethnic minorities it is only 20 per cent of the more integrated families that responded to higher immigrant concentrations by opting out. These results lend support to the idea that native flight from immigrant-dense schools and ethnic segregation across schools are influenced by Danes and immigrants having differing behaviour. A Nordic survey-based study (Brattbakk et al. 2013) among natives leaving multi-ethnic neighbourhoods showed that 17 per cent in Helsinki, 19 per cent in Stockholm and 45 per cent in Oslo stated that too many ethnic minorities in the schools was an important reason for leaving. As this study showed, the role of schools in ethnic segregation is very context dependent and differs greatly across countries and cities. In Norway it could be important that choice of school is more restricted to local schools than in the other countries.

The reasons for effects of local schools on dissatisfaction and moves are, however, not quite clear. Many bilingual pupils in a school could result in a greater strain on resources leading to a lower quality of education. Problems in schools can, however, also mirror social problems in the neighbourhood. If this is the dominant cause, the effects of school problems are connected to the effects of social problems in a neighbourhood. A qualitative study among natives in multi-ethnic neighbourhoods in Finland (Komulainen 2012) on the importance of schools thus found that families' decisions to move mostly were more determined by other things than dissatisfaction with schools or ethnic mix in the neighbourhood.

Causes of selective migration among ethnic minorities

As has been described in Chapters 2–4, more research has been done on causes of diverging neighbourhood choices and moving behaviour among ethnic minorities. In the literature, two different main explanations have been put forward for why immigrants' choice of residence might differ from natives' choice of residence. The first line of explanations focuses on the options and constraints immigrants face on the housing market and the importance of the spatial distribution of different kinds of housing across neighbourhoods in the cities. This will be treated in detail in Chapters 6 and 7. Research about this has found that the availability of housing supply preferred by and obtainable for ethnic minorities, and the way this supply is distributed across neighbourhoods in a city, has a fundamental importance for where ethnic minorities settle. The likelihood for immigrants of gaining access to a particular type of neighbourhood is thus dependent on the quantity and quality of destination opportunities on local housing markets, especially the supply of rented housing. Some researchers have pointed to this as a special kind of explanation for selective migration and ethnic segregation called *the housing availability model* (see Chapter 7).

The second line of explanations interrogates if immigrants, for different reasons, have special preferences for tenure (Chapter 2) and for where to settle in cities. It especially concerns whether new immigrants prefer to settle in neighbourhoods with an ethnic network they can get support from – so-called preferences for *ethnic enclaves* (Chapter 3) – and whether such preferences weaken in pace with cultural, social and economic integration/assimilation leading to a dispersal in the city – so-called *spatial assimilation* (Chapter 4). The connection between preferences for the ethnic composition of neighbourhoods and actual moves was tested in a Danish study (Skifter Andersen 2015a and 2006). Results from this study have also been used in Chapter 3. It showed that both preferences for living close to an ethnic network and for settling in multi-ethnic neighbourhoods were highly significant in explaining actual moves to multi-ethnic neighbourhoods, with more than 40 per cent immigrants among the residents. It also showed a strong importance for actual moves of social integration and resources among ethnic minorities.

The implication of this is that it should be expected that some ethnic minorities tend to be more attracted to multi-ethnic neighbourhoods than to similar neighbourhoods with fewer ethnic minorities, producing a kind of selective migration that could be called *ethnic attraction*. It must be assumed, however, that there is a strong connection between white avoidance and over-representation of ethnic minorities among in-movers to multi-ethnic neighbourhoods. When the native majority to some extent rejects to move into such neighbourhoods, it will be easier for ethnic minorities to obtain housing there. White avoidance and ethnic attraction are thus linked phenomena.

Important for the development of multi-ethnic neighbourhoods is also to what extent ethnic minorities more often tend to stay put in these areas compared to similar neighbourhoods. If they less often move away than they do from similar places, one could talk about *ethnic retention*. Following the spatial assimilation theory, ethnic minorities should move away from multi-ethnic neighbourhoods in pace with their integration in society. Alternative theories have, however, been put forward saying that the advantages for ethnic minorities living in neighbourhoods with a strong ethnic network are so important that they exceed the benefits of spatial assimilation (see Chapter 3). Moreover, restricted options for ethnic minorities on the housing market could force them to remain. Boschman et al. (2017) thus found in the Netherlands that ethnic minorities and low-income household were less likely than natives to realise their moving desires. If white flight means that the native majority more often leave multi-ethnic neighbourhoods, the simple reason could then be that they are more able than ethnic minorities to realise their wishes to change their housing situation. Schaake et al. (2010) found in the same country that ethnic minorities were more likely than native Dutch to move within multi-ethnic neighbourhoods, and less likely to move away from them, when the share of Non-Western minorities in those neighbourhoods increases.

Some studies have shown that ethnic minorities do not always want to live in neighbourhoods with high concentrations of ethnic minorities. The American studies of Clark (2002) and Farley et al. (1997) both concluded that African-Americans preferred to live in neighbourhoods with a mix of ethnic groups, but not neighbourhoods dominated by white Americans. The Danish study reported in Chapter 3 also showed that very few minorities wanted to live in neighbourhoods where the native population was a minority, but that 30 per cent preferred an equal mix of natives and immigrants.

Finally, ethnic minorities' inclination to leave multi-ethnic neighbourhoods might be influenced by the same kind of neighbourhood problems and bad reputation as the native majority. Phillips' (2008) study among British Muslims showed that especially concerns over crime and safety fostered preferences for more mixed neighbourhoods, leading some to actually move into more mixed areas. A Danish study by Skifter Andersen (2006) compared the reasons for why, respectively, Danes and ethnic minorities moved out of three large multi-ethnic neighbourhoods. Both groups left primarily for housing reasons, which were most important for ethnic minorities. There were only small differences between the two groups in motives for leaving connected to the neighbourhood. For both groups, bad reputation was the most important reason, followed by antisocial behaviour. Only 6 per cent of moving Danes complained about the ethnic composition of the neighbourhoods, while 11 per cent of moving ethnic minorities did. These results could, however, to some extent be influenced by major differences between the age and household composition of the two groups. The Danes living in these neighbourhoods are younger and more often single, who have a different view of neighbourhood qualities than families with children.

A Danish study of selective migration

The purpose of this study (described in Skifter Andersen 2017) was to estimate which kind of selective migration has been the major force in creating multi-ethnic neighbourhoods in Denmark. Selective migration among Danes and ethnic minorities was examined, and the sizes of different kinds of migration were estimated, based on data on moves in Denmark in all the years from 1985 to 2008. Four kinds of special moving behaviour are examined:

- *White flight*: Danes more often move away from multi-ethnic neighbourhoods than they normally would from similar neighbourhoods with an average ethnic composition
- *White avoidance*: Danes less often move into multi-ethnic neighbourhoods than they normally would to similar neighbourhoods
- *Ethnic attraction*: Ethnic minorities more often move to multi-ethnic neighbourhoods than they normally would to similar neighbourhoods
- *Ethnic retention*: Ethnic minorities less often move away from multi-ethnic neighbourhoods than they normally would from similar neighbourhoods.

This section contains two kinds of investigations. The first kind analyses whether selective migration behaviour can be identified by using logistic regressions of factors determining when Danes and immigrants move to and from neighbourhoods with a different proportion of ethnic minorities among residents. The second makes an effort to estimate the magnitude of these movements to show their importance for the creation and maintenance of multi-ethnic neighbourhoods in Denmark.

The study is based on a more extensive and richer data set than previous studies. Moreover, new statistical methods have been employed. Earlier European studies have mostly used simple statistics on moves in and out of neighbourhoods. In one of the studies (Andersson 2013), logistic regression has been used to examine moves into neighbourhoods with an increasing ethnic concentration when compared to moving into other neighbourhoods. In general it has been more common in studies of mobility to make statistical models on out-mobility, where it has been possible to use both neighbourhood and personal variables. In studies of choice of neighbourhoods, the used method has been to compare selected neighbourhoods with other neighbourhoods (Hedman et al. 2011), making it impossible to use personal variables. This study uses another method to examine in-moves. Instead of comparing neighbourhoods, a comparison of in-movers is used to examine if neighbourhood conditions effect the composition of in-movers. This makes it possible to include personal variables in the analyses, making them more reliable.

In the first part of the section, a test for selective migration for Danes and Non-Western immigrants is made by help of logistic regressions. Individual moves away from neighbourhoods every year in the period are analysed by regressions for both Danes and immigrants, controlling for data on individuals and their household, including data on changes in their situation the current year, data on the neighbourhood they leave concerning proportion of ethnic minorities, residents without work and housing, and data on the municipality, its urbanisation and proportion of immigrants in social housing. Moreover, the time period is a variable. The pivotal variables in the analyses are the proportion of ethnic minorities in the neighbourhood and the social composition measured by proportion of residents in employment. The last variable makes it possible to test if white flight is present also when controlling for the social composition of the neighbourhood, testing the so-called racial proxy effect.

In-migration to neighbourhoods is analysed by logistic regression based on a data set consisting of all moving Danes and ethnic minorities using the same independent variables except for the data on changes in household situation. The logic in the models is to test for if an in-mover is either a Dane or belonging to an ethnic minority, depending on the proportion of ethnic minorities and other neighbourhood variables and controlled for other individual data than ethnic group, etc. The data and methods used are described in more detail in Skifter Andersen (2017).

In the second part, an effort is made to estimate the magnitude of these movements to show their importance for the spatial concentration of Non-Western ethnic minorities in Denmark. The method is to use regression models to estimate expected moving patterns of both immigrants and Danes, which are not influenced by the ethnic composition of neighbourhoods. The data used are aggregated on neighbourhoods/years based on the selected individuals 15+ years, not living with parents. Two data sets were constructed on the neighbourhood level; one with Danes alone and one with ethnic minorities. Two regressions have been made on the neighbourhood data with Danes, with dependent variables defined as the proportion moving respectively in and out of neighbourhoods during the last year. And two other similar regressions were made on neighbourhood data with ethnic minorities. The independent variables concern the social and demographic composition of the neighbourhood, housing tenures, the municipality and the time period. The ethnic composition is not included. The statistical models are used to calculate the number of moves for both Danes and immigrants to and from every neighbourhood/year, which should be expected independently of their ethnic composition. These expected moves are then compared with actual moves to and from neighbourhoods with different proportions of ethnic minorities to estimate the size of the four kinds of selective moves. The validity of the method depends on the extent to which the connection between ethnic composition and other neighbourhood variables are decisive for the results. This problem is more crucial for immigrants, as a larger share of them are living in multi-ethnic neighbourhoods.

Moves out of multi-ethnic neighbourhoods

Figure 5.1 shows the average proportion of ethnic minorities and Danes, respectively, who moved away within a year from neighbourhoods with different degrees of concentration of ethnic minorities.

There is a clear tendency in the figure. In neighbourhoods with few ethnic minorities, which constitute by far the major part of the population, ethnic minorities more often move away from the neighbourhood than Danes. However, even in neighbourhoods with 20–29 per cent ethnic minorities, Danes more often leave than ethnic minorities, and the difference between the two groups rises with an increasing concentration of ethnic minorities. In the most immigrant-dense neighbourhoods, with more than 70 per cent ethnic minorities, the mobility increases markedly for both Danes and ethnic minorities.

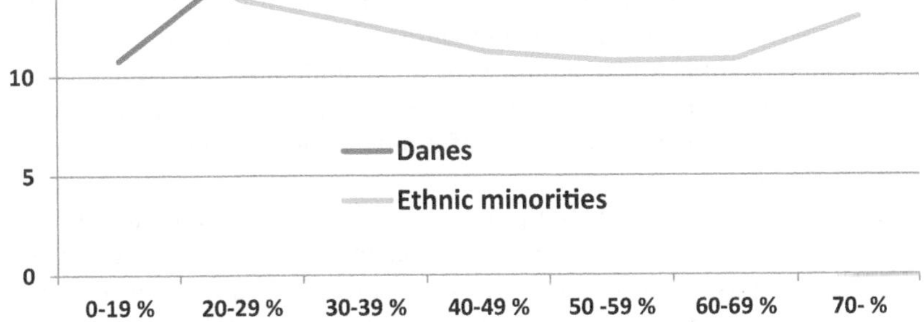

Figure 5.1 Average proportion (%) of ethnic minorities and Danes, respectively, who moved away within a year from neighbourhoods with different percentages of Non-Western ethnic minorities among residents, 1985–2008

Source: Skifter Andersen H. 2017

These average tendencies might, however, be caused by other differences between the neighbourhoods; among others that the social composition might be different. Some statistical analyses (logistic regression) were made to control for such differences and for differences among residents moving away. The dependent variable was moving away from the neighbourhood or not. The independent variables used were: 1. *Neighbourhood variables*: Proportion of ethnic minorities, Neighbourhood size: number of residents, Proportion of residents in employment: Position in a neighbourhood hierarchy divided into quintiles with increasing average employment and Proportion of social housing in neighbourhood, 2. *Person/household variables*: Ethnic group, Sex, Age group, Family type, Housing space per person before move in sq. m., Employment and Household income level: National quintile, 3. *Actual changes for persons/households*: Moving in with a partner, Getting divorced, Getting employment, Increased income (moving up to a higher quintile) and Getting a new job, 4. *Municipality variables*: Proportion of ethnic minorities in municipality, Proportion of residents in social housing, Urbanisation and 5. *Time period*: 1985–87, 1988–92, 1993–97, 1998–02, 2003–06, 2007–08.

Three analyses have been carried through for all movers, Danes and ethnic minorities. Table 5.1 shows the odds ratios for two of the neighbourhood variables, proportion of ethnic minorities and proportion in employment. Moreover, the ethnic group of the moving person is showed. The results for control variables are shown in Skifter Andersen (2017).

The results confirm selective behaviour among Danes and immigrants. The average mobility out of multi-ethnic neighbourhoods for all residents does not seem to be much larger in neighbourhoods with many minorities than for all other neighbourhoods (first column) when controlling for other variables. The social composition of neighbourhoods has some importance, as mobility is higher in all neighbourhoods with higher unemployment than neighbourhood quintile 5 with the highest employment rate. The differences between the first four quintiles are, however, quite small.

For Danes the probability of out-mobility is clearly increasing with proportion of immigrants and the odds ratio is up to 40 per cent higher in the second-most immigrant-dense neighbourhoods. For ethnic minorities there in general is a little lower mobility in all neighbourhoods with more than 20 per cent ethnic minorities, but there is only a weak connection between mobility and increasing proportion of ethnic minorities in the multi-ethnic neighbourhoods.

The results indicate that racial proxy effects for out-migration are weak. For both Danes and ethnic minorities, the effect of the social composition of the neighbourhoods is small; furthermore, small differences are found between the first four employment quintiles cover 80 per cent of the population.

The lowest section in the table shows some differences between different ethnic groups. In general out-mobility is higher among ethnic minorities than among Danes and especially among immigrants from Other Africa. Mobility is much lower among East European immigrants.

Table 5.1 Results on neighbourhood variables from logistic regression (odds ratios) of moving away from neighbourhoods within a certain year, 1985–2008, and individual data on ethnic group

	All	*All Danes*	*Ethnic minorities*
Proportion ethnic minorities (ref. 0–19%)			
20–29%	1.02	1.05	.94
30–39%	1.02	1.09	.91
40–49%	1.04	1.19	.88
50–59%	1.04	1.33	.87
60–69 %		1.41	.86
70– %	.93	1.34	.94
Proportion in employment, neighbourhood quintiles (ref. Quintile 5)			
Quintile 1	1.16	1.16	1.03
Quintile 2	1.23	1.23	1.05
Quintile 3	1.21	1.22	1.03
Quintile 4	1.18	1.18	1.05
Ethnic group (ref Denmark)			
Eastern Europe	1.1		0.89
Middle East and North Africa	1.18		1.08
Other Africa	1.36		1.25
Other Asia	1.17		
Nagelkerke R^2	.261	.263	.207

Source: Skifter Andersen H. 2017

Note: All shown results are significant within the 5 per cent limit.

Moves into multi-ethnic neighbourhoods

As can be seen from Figure 5.2, there has on average been a straight forward negative connection between the proportion of Danes among in-movers and the proportion of ethnic minorities among residents in the neighbourhoods. There has been nearly the same connection in all periods since 1985.

To make a more robust measure of the effect of the ethnic composition of the neighbourhoods on the composition of in-movers, statistical analyses (logistic regressions) were conducted to control for the influence of other factors. The analyses are made on a selection of all moving Danes and immigrants in the database. The dependent variable is whether in-movers to a neighbourhood are Danes or Non-Western ethnic minorities, depending on neighbourhood variables. The used independent variables are the same as used in the analysis of white flight, except for that some of the variables showing changes in household and job situation are omitted. A variable for the ethnic composition of ethnic minorities in the neighbourhood is added.

Table 5.2 shows the odds ratios for some of the neighbourhood variables.

It can be seen that the proportion of ethnic minorities in the neighbourhoods has a dramatic effect on the composition of in-movers. Even in neighbourhoods with only 20–29 per cent ethnic minorities, the odds ratio for Danes among in-movers is half of that of neighbourhoods with less than 20 per cent. The odds ratio

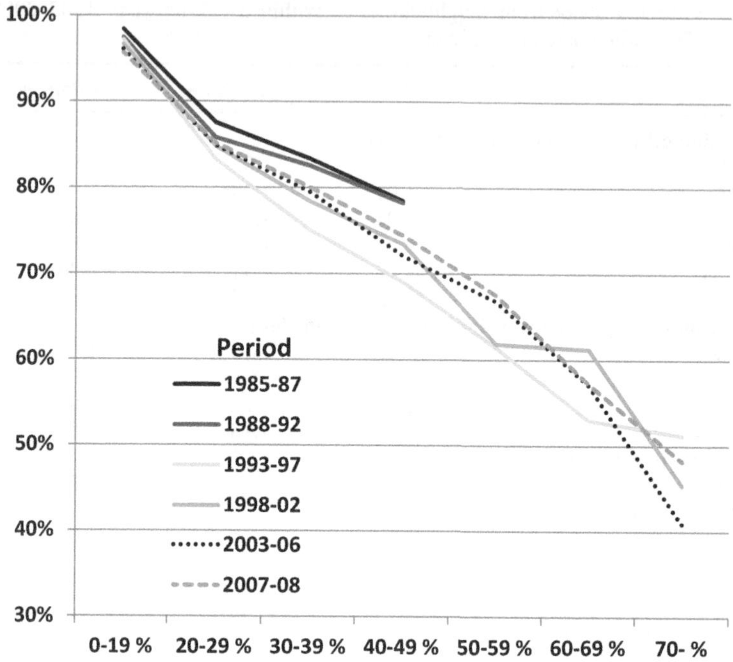

Figure 5.2 The average proportion of Danes among in-movers to neighbourhoods with different percentages of Non-Western ethnic minorities among residents, 1985–2008

Source: Skifter Andersen H. 2017

decreases further with increased concentration of ethnic minorities and is very small in the most immigrant-dense neighbourhoods. Instead the probability of ethnic minorities among in-movers is sharply increasing. The study (reported in Skifter Andersen 2017) also showed that the income of in-moving Danes decreased with an increasing proportion of ethnic minorities in the neighbourhood.

Table 5.2, moreover, shows that the social composition of the neighbourhoods has some importance for whether Danes move in. The probability that an in-mover is a Dane is much smaller for neighbourhoods in the highest unemployment quintile 1, and the probability increases with higher quintiles contrary to what was found for out-mobility in Table 5.1. Racial proxy effects are thus notable among in-moving Danes, but they do not at all eliminate the effects of the ethnic composition of neighbourhoods. The reverse pattern is consequently found among ethnic minorities. They are more likely to move into neighbourhoods with many unemployed residents.

The national background of ethnic minorities in the neighbourhoods is also important. Three variables are used: the proportion of Non-Western residents from respectively the Middle East and North Africa, Other Africa and Other Asia. The reference is residents from Eastern Europe. Especially many residents in a neighbourhood from the Muslim countries in the Middle East and North Africa

Table 5.2 Neighbourhood factors that influence whether in-movers 15+ years old are Danes or Non-Western ethnic minorities, 1985–2008 (odds ratios from logistic regression)

	Danes	*Ethnic minorities*
Prop. of ethnic min. in nbh. (ref. 0–19%)		
20–29%	.54	1.86
30–39%	.45	2.24
40–49%	.36	2.77
50–59%	.30	3.29
60–69%	.25	4.06
70– %	.14	7.34
Employment rate in nbh. (ref. Quintile 5)		
Quintile 1	.57	1.77
Quintile 2	.65	1.53
Quintile 3	.73	1.37
Quintile 4	.83	1.21
Ethnic composition of ethnic minorities in neighbourhood (ref. East Europeans)		
Prop. from Middle East and North Africa	.59	1.71
Prop. from Other Africa	.77	1.29
Prop. from Other Asia	.59	1.68
Nagelkerke R^2	.249	.249

Source: Skifter Andersen H. 2017

Note: All shown results are significant within the 5 per cent limit.

results in fewer Danes among in-movers, but also when there are many from other Asian countries. Immigrants from Eastern Europe have the smallest effect.

Estimating the extent of the different processes of segregation

In this section, we will try to estimate the magnitude of the four processes:

- *White flight*: Danes more often move away from multi-ethnic neighbourhoods than they normally would from similar neighbourhoods of the same character, but with an average proportion of ethnic minorities
- *White avoidance*: Danes less often move to multi-ethnic neighbourhoods than they normally would to similar neighbourhoods of the same character, but with an average proportion of ethnic minorities
- *Ethnic attraction*: Ethnic minorities more often move to multi-ethnic neighbourhoods than they normally would to similar neighbourhoods of the same character, but with an average proportion of ethnic minorities. This could either be because they have preferences for multi-ethnic neighbourhoods (enclaves) or because it is difficult for them to get access to other housing
- *Ethnic retention*: Ethnic minorities less often move away from multi-ethnic neighbourhoods than they normally would from similar neighbourhoods of the same character, but with an average proportion of ethnic minorities. This might be due to preferences for enclaves or because they have few other options.

The problem, which has to be solved, is to make an estimate of 'typical' in- and out-moves that are independent of the ethnic composition of neighbourhoods, only depending on all other characteristics of the neighbourhoods.

The expected typical moves to and from each neighbourhood are estimated by means of linear regression of the connection between different neighbourhood characteristics and actual moves in the 9,000 neighbourhoods for each of the years 1986 to 2008, in all about 216,000 observations. The data on the social and demographic composition of neighbourhoods are determined by aggregating data on all individuals more than 15 years old, who do not live with parents after moving. Two data sets were constructed on the neighbourhood level; one with Danes alone and one with ethnic minorities. Two regressions have been conducted, with dependent variables defined as the proportion of Danes moving respectively in and out of neighbourhoods, based on neighbourhood data on Danes. And two other similar regressions were conducted on neighbourhood data with ethnic minorities. The independent variables used in the models are: 1. Data on the neighbourhoods: composition of residents regarding age, family situation, employment and income and tenure composition of housing, 2. Urbanisation – city size, 3. Data on municipalities: moves of Danes and ethnic minorities inside and between neighbourhoods, and 4. Time period. Data on the ethnic composition of the neighbourhoods is not included, as we wanted to estimate expected moves that were not influenced by this. A more detailed description of the data and models can be found in Skifter Andersen (2017).

By means of the regression models, the expected number of moves to and from every neighbourhood in every year is calculated as a percentage of the population. The contributions from the four processes to the creation and maintenance of concentration of ethnic minorities in the neighbourhoods are then calculated as:

White flight	=	Actual moves of Danes out of the neighbourhoods minus expected moves
White avoidance	=	Expected in-moves of Danes minus actual in-moves
Ethnic attraction	=	Actual in-moves of ethnic minorities minus expected in-moves
Ethnic retention	=	Expected moves out of ethnic minorities minus actual moves out

The results are shown in Figure 5.3.

It is seen that white avoidance increases strongly with the increasing proportion of ethnic minorities in the neighbourhoods and it is the most important of the four processes in the most immigrant-dense neighbourhoods. When Danes do not move in, it is obvious that the in-movers will be ethnic minorities unless vacant dwellings arise. Ethnic attraction is also quite strong. It is, however, not possible from these data to tell whether it is because ethnic minorities are just 'filling the gaps' when Danes don't move in, or whether it is because they have strong preferences for these neighbourhoods. It must, however, be assumed that it is difficult for ethnic minorities to outcompete Danes in multi-ethnic neighbourhoods, as

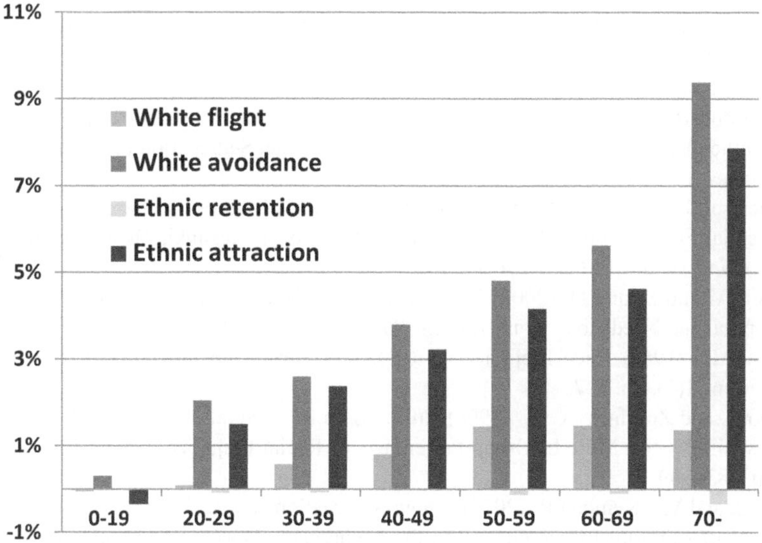

Figure 5.3 Calculated effects of the four processes of selective migration in neighbourhoods with different proportion of Non-Western ethnic minorities among residents, 1985–2008. Percentage of population in neighbourhoods per year

Source: Skifter Andersen H. 2017

these almost exclusively consist of social housing, where dwellings are allocated from waiting lists.

White flight also has some importance that increases with an increasing proportion of ethnic minorities until 50 per cent. It is a little smaller in the most minority-dense neighbourhoods. An explanation could be that Danes left behind in these neighbourhoods are mostly low-income families with no other options.

Ethnic retention, the phenomenon that ethnic minorities tend to remain longer in multi-ethnic neighbourhoods, seems to have little importance in the Danish case. It is a little unexpected, compared with the analysis in Table 5.1, that ethnic minorities somewhat more often than should be expected move out of neighbourhoods with higher proportions of ethnic minorities. There is some uncertainty connected with the method used and the figures are small, but it can be concluded that ethnic retention is not an important factor in the spatial concentration of ethnic minorities in Denmark.

References

Abascal M. and Baldassari D. (2015) Love Thy Neighbor? Ethno-racial Diversity and Trust Re-examined, *American Journal of Sociology*, 121(3), pp. 722–782.

Ærø T. (2006) Residential Choice from a Lifestyle Perspective, *Housing, Theory and Society* 23, pp. 109–130.
Andersson R. (2013) Reproducing and Reshaping Ethnic Residential Segregation in Stockholm: The Role of Selective Migration Moves, *Geografiska Annaler: Series B, Human Geography*, 95(2), pp. 163–187.
Andersson R. and Bramå Å. (2006) Selective Migration in Swedish Distressed Neighbourhoods: Can Area-based Urban Policies Counteract Segregation Processes? *Housing Studies*, 19(4), pp. 517–539.
Andersson R., Brattbakk I. and Vaattovaara M. (2017) Natives' Opinions on Ethnic Residential Segregation and Neighbourhood Diversity in Helsinki, Oslo and Stockholm, *Housing Studies*, 32(4), pp. 491–516.
Basolo V. and Strong D. (2002) Understanding the Neighborhood: From Residents' Perception and Needs to Action, *Housing Policy Debate*, 13(1), pp. 83–105.
Blumer H. (1958) Race Prejudice as a Sense of Group Position, *Pacific Sociological Review*, 1(1), pp. 3–7.
Bobo L. and Zubrinsky C. L. (1996) Attitudes on Residential Integration: Perceived Status Differences, Mere In-Group Preference, or Racial Prejudice? *Social Forces*, 74(3), pp. 883–909.
Bolt G. and Van Kempen R. (2010) Ethnic Segregation and Residential Mobility: Relocations of Minority Ethnic Groups in the Netherland, *Journal of Ethnic and Migration Studies*, 36(2), pp. 333–354.
Bolt G., Van Kempen R. and Van Ham M. (2008) Minority Ethnic Groups in the Dutch Housing Market: Spatial Segregation, Relocation Dynamics and Housing Policy, *Urban Studies*, 45(7), pp. 1359–1384.
Boschman S., Kleinhans R. and van Ham M. (2017) Ethnic Differences in Realising Desires to Leave Urban Neighbourhoods, *Journal of Housing and the Built Environment*, 32, pp. 495–512.
Boterman W. R. (2012) Dealing with Diversity: Middle-class Family Households and the Issue of 'Black' and 'White' Schools in Amsterdam, *Urban Studies*, 50(6), pp. 1130–1147.
Brattbakk I., Andersson R. and Dhalman H. (2013) The Role of the School for Residential Mobility and Ethnic Residential Segregation in Nordic Capital Regions. *Extended abstract for the ENHR-conference in Tarragona*, June 19–22, 2013.
Burgess S., Wilson D. and Lupton R. (2005) Parallel Lives? Ethnic Segregation in Schools and Neighbourhoods, *Urban Studies*, 42(7), pp. 1027–1056.
Burrows R. and Rhodes D. (1999) *Unpopular Places? Area Disadvantage and the Geography of Misery in England*. Bristol: The Policy Press.
Cascio E. U. and Lewis E. G. (2012) Cracks in the Melting Pot: Immigration, School Choice and Segregation, *The American Economic Journal: Economic Policy*, 4(3), pp. 91–117.
Clark W. A. V. (2002) Ethnic Preferences and Ethnic Perceptions in Multi-Ethnic Settings, *Urban Geography*, 23(3), pp. 237–256.
Clark W. A. V. and Ledwith V. (2005) *Mobility, Housing Stress and Neighbourhood Contexts: Evidence from Los Angeles*. On-line Working Paper Series, University of California, California Center for Population Research.
Clark W. A. V. and Maas R. (2012) Schools, Neighbourhoods and Selection: Outcomes Across Metropolitan Los Angeles, *Population Research and Policy* Review, 31(3), pp. 339–360.

Costa Pinho T. (2000) Residential Contexts of Social Exclusion: Images and Identities, *Paper for ENHR Conference, Gävle, 2000.*

Crowder K., Hall M. and Tolnay S. E. (2011) Neighborhood Immigration and Native Out-migration, *American Sociological Review*, 76(1), pp. 25–47.

Crowder K. and South S. J. (2008) Spatial Dynamics of White Flight: The Effects of Local and Extralocal Racial Conditions on Neighborhood Out-migration, *American Sociological Review*, 73(5), pp. 792–812.

Dekker K. (2013) Testing the Racial Proxy Hypothesis. What is it that Residents Don't Like about Their Neighbourhood? In van Hamm M., Manley D., Bailey N., Simpson L. and Maclennan D. *Neighbourhood Dynamics. New Insights for Neighbourhood Effects Research*. Dordrecht: Springer.

Dekker K., Musterd S. and Van Kempen R. (2007) Explaining Differentials in Housing and Neighbourhood Satisfaction in Post-WWII Large Housing Estates in European Cities, *Paper for ENHR Conference 2007, Rotterdam.*

Dekker K., de Vos S., Musterd S. and Van Kempen R. (2011) Residential Satisfaction in Housing Estates in European Cities, *Housing Studies*, 26(4), pp. 479–499.

Ellen I. G. (2000) Race-based Neighbourhood Projection: A Proposed Framework for Understanding New Data on Racial Integration, *Urban Studies*, 36(9), pp. 1513–1533.

Farley R., Fielding E. L. and Krysan M. (1997) The Residential Preferences of Blacks and Whites: A Four-metropolis Analysis, *Housing Policy Debate*, 8(4), pp. 763–800.

Farley R., Steeh C. and Krysan M. (1994) Stereotypes and Segregation: Neighborhoods in the Detroit Area, *American Journal of Sociology*, 100(3), pp. 750–780.

Feijten P. and van Ham M. (2009) Neighbourhood Change. Reason to Leave? *Urban Studies*, 46, pp. 2103–2122.

Galster G. (1990) Neighborhood Racial Change, Segregationist Sentiments, and Affirmative Marketing Policies, *Journal of Urban Economics*, 27(3), pp. 344–361.

Gramberg P. (1998) School Segregation: The Case of Amsterdam, *Urban Studies*, 35(3), pp. 547–564.

De Groot C., Manting D. and Mulder C. H. (2007) Intentions to Move and Actual Moving Behaviour in the Netherlands, *Paper for ENHR Conference 2007, Rotterdam.*

Harris R. H. (2001) Why Are Whites and Blacks Averse to Black Neighbors? *Social Science Research*, 30, pp. 100–116.

Havekes E., Coenders M. and van der Lippe T. (2014) The Wish to Leave Ethnically Concentrated Neighbourhoods: The Role of Perceived Social Cohesion and Interethnic Attitudes, *Housing Studies*, 29(6), pp. 823–842.

Hedman L., van Ham M. and Manley D. (2011) Neighbourhood Choice and Neighbourhood Reproduction, *Environment and Planning A*, 43, pp. 1381–1399.

Johnston R., Burgess S., Wilson D. and Harris R. (2006) School and Residential Ethnic Segregation: An Analysis of Variations across England's Local Education Authorities, *Regional Studies*, 40(9), pp. 973–990.

Kaufmann E. and Harris G. (2015) 'White Flight' or Positive Contact? Local Diversity and Attitudes to Immigration in Britain, *Comparative Political Studies*, 48(12), pp. 1563–1590.

Kearns A. and Parkes A. (2003) Living in and Leaving Poor Neighbourhood Conditions in England, *Housing Studies*, 18(6), pp. 827–852.

Komulainen S. (2012) 'White Flight' in Finland? A Qualitative Study into Finnish-born Families' Housing and School Choices in Turku, *Finnish Yearbook of Population Research*, XLVII, pp. 51–64.

Koopman M. J. (2007) Looking from the Inside Out or from the Outside In: Differences and Associations between Neighbourhood Push and Pull. *Paper for ENHR Conference 2007, Rotterdam.*

Ledwith V. and Clark W. A. V. (2007) The Effect of the Residential Mosaic and "White Flight" in Public School Composition: Evidence from Los Angeles County, *Urban Geography*, 28(2), pp. 160–180.

Lee B. A., Oreposa R. S. and Kanan J. W. (1994) Neighbourhood Context and Residential Mobility, *Demography*, 31(2), pp. 249–270.

Lu M. (1998) Analysing Migration Decision Making: Relationship between Residential Satisfaction, Mobility Intentions and Moving Behaviour, *Environment and Planning A*, 30, pp. 1473–1495.

Magnusson L. and Özüekren A. Ş. (2002) The Housing Careers of Turkish Households in Middle-sized Swedish Municipalities, *Housing Studies*, 17(3), pp. 465–486.

Pais J. F., South S. and Crowder K. (2009) White Flight Revisited: A Multiethnic Perspective on Neighborhood Out-migration, *Population Research Policy Review*, 28(3), pp. 321–346.

Parkes A., Kearns A. and Atkinson R. (2002) What Makes People Dissatisfied with Their Neighbourhoods? *Urban Studies*, 39, pp. 2413–2438.

Permentier M., Bolt G. and van Ham M. (2011) Determinants of Neighbourhood Satisfaction and Perception of Neighbourhood Reputation, *Urban Studies*, 48(5), pp. 977–996.

Permentier M., Van Ham M. and Bolt G. (2007) Behavioural Responses to Neighbourhood Reputations, *Journal of Housing and the Built Environment*, 22(2), pp. 199–213.

Pettigrew T. F. and Tropp L. R. (2006) A Meta-analytic Test of Intergroup Contact Theory, *Journal of Personality and Social Psychology*, 90(5), pp. 751–783.

Phillips D. (2008) The Problem with Segregation: Exploring the Racialisation of Space in Northern Pennine Towns. In Dwyer C. and Bressey C. *New Geographies of Race and Racism*, pp. 179–192. Aldershot, UK: Ashgate.

Putnam R. D. (2007) E Pluribus Unum: Diversity and Community in the Twenty-first Century the 2006 Johan Skytte Prize Lecture, *Scandinavian Political Studies*, 30(2), pp. 137–174.

Quillian L. (2002) Why is Black – White Residential Segregation So Persistent? Evidence on Three Theories from Migration Data, *Social Science Research*, 31, pp. 197–202.

Rangvid B. S. (2009) School Choice, Universal Vouchers and Native Flight from Local Schools, *European Sociological Review*, 26(3), pp. 319–335.

Rijpers B. and Smeets J. (1998) Housing Challenge: Managing Neighbourhoods Image. *Paper to ENHR conference, Cardiff 7–11. September 1998.*

Schaake K., Burgers J. and Mulder C. H. (2014) Ethnicity, Education and Income, and Residential Mobility between Neighbourhoods, *Journal of Ethnic and Migration Studies*, 40(4), pp. 512–527.

Schaake K., Burgers J. and Mulder C. H. (2010) Ethnicity at the Individual and Neighborhood Level as an Explanation for Moving Out of the Neighborhood, *Population Research Policy Review*, 29(4), pp. 593–608.

Semyonov M. and Glikman A. (2009) Ethnic Residential Segregation, Social Contacts, and Anti-Minority Attitudes in European Societies, *European Sociological Review*, 25(6), pp. 693–708.

Semyonov M., Glikman A. and Krusan M. (2007) Europeans' Preference for Ethnic Residential Homogeneity: Cross-National Analysis of Response to Neighborhood Ethnic, *Social Problems*, 54(4), pp. 434–453.

Skifter Andersen H. (2017) Selective Moving Behaviour in Ethnic Neighbourhoods: White Flight, White Avoidance, Ethnic Attraction or Ethnic Retention? *Housing Studies*, 32(3), pp. 296–318.

Skifter Andersen H. (2015a) Explanations for Special Neighbourhood Preferences among Ethnic Minorities, *Housing, Theory and Society*, 32(2), pp. 196–217.

Skifter Andersen H. (2015b) *Indvandring, integration og etnisk segregation* (Immigration, Integration and Ethnic Segregation). Copenhagen: Danish Building Research Institute at Aalborg University.

Skifter Andersen H. (2008) Why Do Residents Want to Leave Deprived Neighbourhoods? – The Importance of Residents' Subjective Evaluations of Their Neighbourhood and Its Reputation, *Journal of Housing and the Built Environment*, 23, pp. 79–101.

Skifter Andersen H. (2006) *Undersøgelse af til- og fraflytningen fra tre multietniske boligområder* (A Study of Reasons for Moves to and from Three Multi-ethnic Neighbourhoods), SBi 2006: 12. Horsholm: Danish Building Research Institute.

Skifter Andersen H. (2002) Excluded Places. On the Interaction between Segregation, Urban Decay and Deprived Neighbourhoods, *Housing, Theory and Society*, 19, pp. 153–169.

Skifter Andersen H., Magnusson Turner L. and Søholt S. (2013) The Special Importance of Housing Policy for the Housing Situation of Ethnic Minorities. Evidence from a Comparison of Four Nordic Countries, *International Journal of Housing Policy*, 13(1), pp. 17–36.

South S. J. and Crowder K. D. (1997) Escaping Distressed Neighborhoods: Individual, Community, and Metropolitan Influences, *American Journal of Sociology*, 102, pp. 1040–1084.

South S. J., Crowder K. and Chavez E. (2005) Migration and Spatial Assimilation among U.S. Latinos: Classical Versus Segmented Trajectories, *Demography*, 42(3), pp. 497–521.

South S. J., Crowder K. and Pais J. (2011) Metropolitan Structure and Neighborhood Attainment: Exploring Intermetropolitan Variation in Racial Residential Segregation, *Demography*, 48(4), pp. 1263–1292.

Stein R. M., Post S. S. and Rinden A. L. (2000) Reconciling Context and Contact Effects on Racial Attitudes, *Political Research Quarterly*, 53(2), pp. 285–303.

Swaroop S. and Kryzan M. (2011) The Determinants of Neighborhood Satisfaction: Racial Proxy Revisited, *Demography*, 48(3), pp. 1203–1229.

Tsfati Y. and Cohen J. (2003) On the Effect of the "Third-person Effect": Perceived Influence of Media Coverage and Residential Mobility Intentions, *Journal of Communication*, 53, pp. 711–727.

van Der Laan Bouma-Doff W. (2007) Involuntary Isolation: Ethnic Preferences and Residential Segregation, *Journal of Urban Affairs*, 29(3), pp. 289–309.

van Ham M. and Clark W. A. V. (2009) Neighbourhood Mobility in Context: Household Moves and Changing Neighbourhoods in the Netherlands, *Environment and Planning A*, 41, pp. 1442–1459.

Van Kempen R. (2003) *Segregation and Housing Conditions of Immigrants in Western European Cities*, Eurex Lecture 7, 13 March. Website: at www.shiva.uniurb.it/eurex/syllabus/lecture7/Lecture7-VanKempen.pdf.

Vilkama K. (2011) Yhteinen kaupunki, eriytyvät kaupunginosat? Kantaväestön ja maahanmuuttajataustaisten asukkaiden alueellinen eriytyminen ja muuttoliike pääkaupunkiseudulla [Shared City, Divided Neighbourhoods? (Residential Segregation and Selective

Migration of the Native and Immigrant Populations in the Helsinki Metropolitan Area). *Tutkimuksia 2011:2*, Helsingin kaupungin tietokeskus, Helsinki.

Wessel T. (2009) Does Diversity in Urban Space Enhance Intergroup Contact and Tolerance? *Geografiska Annaler: Series B, Human Geography*, 91(1), pp. 5–17.

Wright R., Ellis M. and Parks V. (2005) Re-placing Whiteness in Spatial Assimilation Research, *City and Community*, 5(2), pp. 111–135.

Zhang H. (2011) School Desegregation and White Flight Revisited: Analysis from a Metropolitan Perspective, *Urban Geography*, 32(8), pp. 1208–1226.

6 The effects of housing policy and housing markets on housing options and choices of ethnic minorities[1]

Housing policies and housing markets differ extremely across countries. In this chapter is examined how different aspects of housing policy, and the resulting housing markets, affects the housing options and choices of immigrants and ethnic minorities in general. Limitations in the housing options of ethnic minorities is an important cause of ethnic segregation, depending on to what extent housing available for them is concentrated in certain parts of the cities.

Ethnic minorities often have lower incomes. Their housing options therefore to some extent are connected to the options of low-income groups in general. Housing policies and housing markets that make housing more available for families with lower incomes also make it easier for ethnic minorities to obtain good housing. It has been shown, however, that ethnic minorities face other difficulties in the housing market than natives with lower incomes. Moreover some ethnic minorities have, as shown in Chapters 2 and 3, special preferences for housing and neighbourhoods that affect their options and choice of housing. It was thus shown in Chapter 2 that ethnic minorities more often than natives prefer to live in rented housing regardless of their income. The availability of rented housing is thus essential for their housing options. Some of them also have special preferences for living in neighbourhoods with ethnic networks, as shown in Chapter 3. The kind of housing available in such neighbourhoods thus influences their housing preferences and options. Finally, the extent to which the native majority avoid neighbourhoods dominated by ethnic minorities has importance. Housing in such neighbourhoods may be easier to access for ethnic minorities because demand from natives is low.

In the research literature there has been made a distinction between 'dual' and 'unitary' housing markets. In dual markets homeownership is mostly limited to upper income groups, while families with lower incomes are more confined to the rental market. This to a greater extent limits the options of ethnic minorities to rental housing. Another distinction is to divide between unitary and dual rental markets. In the dual markets, social housing is a restricted sector for the poorest families. In some of the research literature, a hypothesis has been put forward that especially these kinds of markets will result in a high concentration of ethnic minorities in social housing. Other research has indicated a general disadvantage for ethnic minorities on the housing market, because of discrimination and lack of social connections to landlords. The extent of these problems depends on how the housing market is regulated.

The discussion will be illuminated by results from a study comparing the housing situation of ethnic minorities in four Nordic countries and how the differences across the countries can be explained by differences in housing policies and housing markets. One of the important conclusions is that there is not a straightforward connection between uneven distribution of income groups across housing tenures and the distribution of ethnic minorities. This is also shown in Chapter 8 where the situation in additional European countries is compared. The housing situation of ethnic minorities thus cannot be explained solely by their lower incomes. Other explanations are important. Some of them are uncovered by looking at the differences in housing markets and policies, including a discussion of the importance of the housing preferences of ethnic minorities described in Chapter 2.

Housing policies and housing markets in Europe

Housing is not something that is produced and distributed by the public sector. Basically housing is delivered by the private housing market, but the state (and local) authorities make corrections to the market to obtain certain goals for the provision of housing (Bengtsson et al. 2006). In fact all European countries have implemented special housing policies to make such market corrections (Doling 1997), but to a very different extent and with different purposes. In all countries actual housing shortages and housing problems for the poor have appeared that have influenced the implementation of housing policies. However, the chosen policies have varied immensely. This is also the case across countries that have nearly the same kind of welfare state system. Three explanations can be formulated for these differences across countries (Skifter Andersen 2012):

- Variation in the conception of the role of the welfare state in general and in particular to what extent housing is a task for the welfare state
- Variation in the conception of to what extent there are 'market failures' in the housing market, which leads housing supply – especially for the poor – to be insufficient or too expensive
- Variation in actual, visible housing problems and to what extent they are accepted. This depends on the actual situation in the countries concerning wealth, income distribution, interest level, land prices, urban structure, etc. The perception of housing problems has changed over time from World War II, when severe housing shortages appeared, to recent years when housing supply to a reasonable extent can meet demands in many countries. The problems also have changed in connection with cyclical changes in the economic conditions of a country.

Differences in housing policy are to a great extent determined by differences in the opinion about what are the duties of the state and to what extent it should produce and distribute services and consumption. The differences between the kinds of welfare systems in different countries normally also are mirrored in housing policy. Three principally different approaches to the role of housing policy in different countries have been indicated (Doling 1997).

The first, which is particularly applicable in some countries in Southern Europe and the US, is that housing is primarily seen as private consumption in line with other consumables. The state only enters when extreme problems are visible in the form of homelessness and strong deterioration of housing.

The second position, which exists in countries like Britain, Belgium, France, Switzerland and partly Germany (Skifter Andersen and Munk 1994), is that housing policy primarily is designed to help vulnerable groups, who are not, by themselves, able to obtain acceptable housing conditions, while the rest of the population has to survive on the general conditions that exist on the housing market.

The third view perceives housing in general as something that is particularly important for the health and welfare of society, and therefore sees it as the state's task to ensure a good supply of housing for all groups in society. It is thus not only housing for vulnerable groups which is supported, but also housing consumption of broad groups of society. It is especially in the Scandinavian countries and in the Netherlands that this political view, to different extents, has been found.

Differences in housing policy may also be due to different opinions that exist on how well the housing market is able to provide the necessary housing supply (Doling 1997). If the housing market is functioning well, housing shortages and poor housing conditions are only an expression of an inability to pay for decent housing among low-income groups, and only individual economic support to such families is needed. There is thus no need for support for housing production and for a special protected social housing sector. In many countries, economists have argued for a stronger shift of subsidies from production to consumption subsidies. In countries like the UK and the United States, the vast majority of subsidies are given as individual support for housing consumption.

Housing policy can be defined as public initiatives which affect the supply, price and quality of dwellings as well as how they are distributed across households. Housing policy is to some extent intertwined with urban policy that influences where and how dwellings are located in space and the qualities of their neighbourhoods.

Social housing

Social housing is in many countries of special importance for ethnic minorities because it is available for families with lower incomes and often is easier for them to access than homeownership and private renting, where rules for access are more blurred and discrimination can occur, as described later in this chapter.

Social housing can be designed in many different ways. The main characteristics of social housing are that:

1 Rents are below market prices,
2 Vacant dwellings are assigned to people in accordance with needs and ability to pay for housing,
3 The properties and their owners are subject to special rules concerning building activity, administration and financial matters and fixing of rents.

Legislation that directly regulates who can get access to dwellings is most often found in tenures which receive supply subsidies, mostly social housing or publicly owned housing. A recent publication made by UNECE (Rosenfeld 2015)

Universal Model	Generalist Model	Residual Model
Allocation principles		
Open to the whole population (subject to registration on the social housing waiting list)	*Vulnerable households *Special groups *Households below defined income thresholds	*Vulnerable households *Special groups (e.g. refugees, asylum seekers, disabled, mentally challenged, substance abusers, etc.)
Objectives		
Enable access to decent housing for all	Respond to difficulties in accessing the housing market	Respond to the population excluded from the housing market

Figure 6.1 Social housing models based on allocation systems

Source: Based on Rosenfeld/UNECE 2015

compares different systems of access to social housing in European countries. The report uses a model proposed by CECODHAS (2007) for classifying social housing allocation models in the EU (Figure 6.1). In the model social housing systems are divided into three groups, which are very much in accordance with the previous description of three different roles of housing policy.

The *Universal allocation model* was designed to make decent and affordable housing available to the whole population regardless of their income. In this model, housing is considered a public responsibility and is provided either through municipal housing companies (e.g. Sweden) or through non-profit organisations (e.g. the Netherlands, Denmark). Overall, countries with universal allocation models have a higher share of the social housing stock in their housing systems (from 18 per cent to 32 per cent). The rents are cost based; allowances and rent guarantees are available for vulnerable groups. In their original form, housing allocations were based on timely subscription on a waiting list and without income limits. In 2007, there were three countries that used this system: Denmark, Sweden and the Netherlands (Ghekiere 2007).

The *Generalist allocation model* assumes that the objectives of housing policy will be predominantly met by the market. Social housing is aimed only at those households for which the market is unable to deliver housing of decent quality at an affordable price. There is a great diversity in type and size of social housing in the models that belong to this category. The percentage of social housing as a part of the housing system is below 19 per cent in this model and often much less. In the generalist model social housing is allocated by housing providers on the basis of a specific set of rules and procedures, following priority criteria and income ceilings. The rents have a fixed ceiling and households benefit from income-based housing allowances covering part of their rent.

The *Residual allocation model* aims to allocate housing to a restricted category of beneficiaries, usually very vulnerable households that are heavily dependent on state benefits (e.g. unemployed, disabled, elderly, single parents). In the European Union, the social housing for the most vulnerable usually relies on the direct allocation of this type of housing by the local authorities on the basis of need (CECODHAS 2007). According to CECODHAS, rents are cost based or set on the basis of income. Housing allowances (or other forms of general social benefits) are available. The cost-based system is not used in the UK and Ireland. One key trend from this model is a high concentration of the poor in social housing, especially in weak housing market areas. In countries where the residual model is prevalent, the social housing tenure presents a very low percentage of the total housing or caters to increasingly poor and vulnerable populations.

According to Rosenfeld (2015), 3 EU countries have been seen as Universal models (Denmark, Sweden and the Netherlands); 13 have used the Generalist models (Austria, Czechia, France, Finland, Belgium, Germany, Poland, Slovenia, Italy, Luxemburg, Greece, Spain and Slovakia). Some of these countries (France, Finland, Belgium, Germany and Slovakia) have also used the Residual model in some of the social housing. Other countries (the UK, Estonia, Ireland, Hungary, Cyprus, Portugal, Bulgaria, Lithuania, Latvia, Romania and Estonia) have only used the Residual model. However, recent changes in policies have changed the position of countries and the social housing stock is increasingly moving towards 'residual' social housing provision. This *residualisation* of social housing results in a lack of social housing, and increased need results in a higher concentration of the most vulnerable population in the available stock. It increases the risk of segregation and leads to the formation of neighbourhoods with high concentrations of poverty and unemployment.

The importance of housing tenures and social segmentation in the housing market

Housing policy instruments are combined in 'packages' for different housing tenures. In each country is defined a limited amount of different tenures, which are subject to certain legislation and sometimes financial support. The design of tenures is of crucial importance for the functioning of the housing market. The establishment of a social housing sector with direct financial support, rent control, regulation of access and special finance is of special importance.

The most important distinction is between owner-occupied and rented housing, but often tenures exist that are a mix between renting and owing. Sometimes only the dwelling is private ownership, while the building is owned in common with other flat owners (owner-occupied flats), or the residents in common are owners of an association that is the actual owner of the property (cooperatives, share holds). There can be different kinds of owners of rented property, such as public authorities, non-profit housing associations or private landlords. All tenures are subject to different kinds of regulation, subsidies and tax rules, which have a strong influence on which households can get access to which tenures. For this reason, the large differences in housing policies across countries also result in big differences in the tenure composition of the housing market (see Chapter 8).

Depending on how tenures are designed, the housing market can be more or less *segmented.* Segmentation of the housing market is a concept that has been used to describe the way different people are allocated to different parts of the housing market (Lindberg and Lindèn 1989) or that different parts of the housing market are designed to meet different kinds of demand (Rothenburg et al. 1991). Segmentation is created when different tenures to a great extent are made available and attractive for different households, for example divided by income and family situation. Segmentation often means that high-income groups are concentrated in certain parts of the housing market, mostly owner-occupied detached housing, while low-income groups mostly reside in poor private rental housing or social housing. Segmentation has mostly been a result of the way subsidies are designed (tax subsidies in owner-occupation is most favourable for high-income groups while only low-income groups can get housing allowances in rental housing) or by the way access to tenures is regulated (sometimes only very low-income groups can get access to subsidised social housing).

In the literature on comparative housing policies across countries, a distinction has been made between dual and unitary housing markets (Kemeny and Lowe 1998; Kemeny 1995). Dual markets are characterised by a high degree of social segmentation and larger income differences between homeowners and renters. This is not only due to potentially higher income inequality but also because owner-occupation is more difficult to obtain for families with middle and low incomes due to high property prices, difficulties in obtaining loans or high interest rates on mortgages. Dual markets can be found in countries both with a high and a low proportion of homeownership, but housing options for low-income groups are more restricted in dual markets.

Another distinction is made between dual and unitary rental markets. In dual rental markets there is a smaller social housing sector, often with a restricted access for only the very poor (see the preceding description of allocation systems in social housing), while unitary rental markets have a stronger social housing sector that competes on more equal terms with private renting and have more open access.

Countries with both a dual market in general and with a dual rental market have the highest segmentation, with large income differences between tenures. If the rental sector in general is small, low-income groups have great difficulties in getting decent housing. In countries with a unitary housing market and unitary rental market, low-income groups have much better chances to find good housing, especially if there is a larger social housing sector.

Ethnic segmentation of housing markets concerns to what extent ethnic minorities are unevenly distributed on housing tenures compared to natives. It should be expected that countries with a high-income segmentation of the housing market also will have a high ethnic segmentation. It is, however, shown in both the Nordic study described subsequently and in Chapter 8 that there is not a straightforward connection between these two kinds of segmentation of housing markets. The lower incomes of ethnic minorities thus cannot explain why they live in other kinds of housing than natives. Other explanations have to be found. One of them is that the selective moving behaviour among the native majority, as described in Chapter 5, is of fundamental importance. Some other possible explanations are described in the following.

Explanations for the deviant housing situation among ethnic minorities compared to the native majority

The analyses made in Chapter 8 show that the housing situation of ethnic minorities differs greatly from the average of the population in all European countries, and that these differences vary greatly across countries depending on their immigration, social inequality and housing markets. Differences between ethnic minorities and the native majority concerning housing have also been illuminated in the research literature (see for example Musterd 2005; Johnston et al. 2002; Finney 2002; Blom and Henriksen 2008; Skifter Andersen 2012). According to the EWSI Analysis: Immigrant Housing in Europe (2016) in 28 EU countries, ethnic minorities across Europe often are generally vulnerable on the housing market, more dependent on private rentals, more likely to be uninformed of their rights and discriminated against. They also face greater obstacles to access public housing or housing benefits and are more likely to live in substandard and poorly connected accommodations, with less space available and at a higher rental cost burden than the national average.

In the following we will try to unravel some of the reasons for why the housing situation of ethnic minorities differs from natives in European countries and the importance of housing policies and housing markets. In connection with the topic of this book, the identification of conditions that decide the access for ethnic minorities to different housing tenures is the most important subject to illuminate. As discussed in Chapter 7 and illuminated in Chapter 8, an important reason for segregation is a combination of high segmentation of ethnic minorities in housing tenures combined with a skewed geographical location of different tenures. If ethnic minorities are referred to specific tenures and this housing is concentrated in certain parts of the cities, ethnic segregation will be strong.

The main reasons that the housing situation of ethnic minorities differs from natives can be assigned to:

- Their lower incomes
- Special housing preferences, and preferences for ethnic neighbourhoods, which sometimes are dominated by certain tenures
- Defective knowledge about how to act in the housing market and missing social relations to decision makers
- Discrimination from landlords, banks and real estate agents
- The extent of housing supply, especially rented housing, that is accessible for ethnic minorities. Especially scope of and access to social housing

The most important factor determining the housing situation of ethnic minorities is their economic situation, which is linked to their social and economic integration in society. Ethnic minorities from third world countries have in general lower incomes than the native population, but the differences vary greatly depending on duration of stay and ethnic group. Some immigrants come from countries with better education and more developed societies, which enable them to achieve faster integration and economic success. As discussed in Chapter 4 some immigrant groups have more difficulties in getting integrated in mainstream society and some are 'integrated into an ethnic underclass' with permanently low incomes. However, there is not a straightforward connection between income segmentation of housing markets and ethnic segmentation. The housing situation of ethnic minorities is thus also influenced by other factors than income. In the following we will go through different kinds of causes that influence housing choices and options of ethnic minorities and how they are affected by housing policies.

Special ethnic preferences for housing and neighbourhood

It has been shown in Chapter 2 that the cultural background and the special position in society for immigrants in some ways can influence their housing preferences and actual housing choice. For some groups special preferences for living in extended families or close to relatives influence the kind of housing they choose.

The options of ethnic minorities in the housing market are very much determined by their potentials for purchasing their own home. Homeownership is a much more accessible kind of housing than renting, and it is therefore important for the possibilities for choosing where to live in a city if one prefers homeownership and is able to obtain it. Economic resources are decisive, but it has been shown that ethnic minorities for different reasons more seldom move into homeownership than the native population. A considerable number of studies in North America and Europe (see Chapter 2) have shown that general household variables like income and family situation cannot alone explain why ethnic minorities in general have a lower homeownership rate than natives. It has also been found that homeownership differs across different ethnic groups and across different countries. Three reasons have been proposed for the difference between natives and especially Non-Western ethnic minorities. The first is that many immigrants still have economic connections to their country of origin, which strain their finances. It could be in the form of investments in property in the homeland or because they have to send remittances to support family there. The second is that purchase of a home presumes expectations of a solid and permanent income and that immigrants often have a general feeling of uncertainty about the future that makes them reluctant to be homeowners.

Furthermore, there might also be a connection between preferences for neighbourhoods, as described in Chapter 3, and choice of tenure. The composition of tenures in some neighbourhoods with low quality or low cost housing, which immigrants could easily access early in their tenure, has to some extent initiated the processes of segregation of ethnic minorities, which has influenced their choice of tenures. Many of the first immigrants settled in older neighbourhoods in cities, dominated by private renting, where ethnic enclaves were established (see

Chapters 3 and 4). Over time, immigrant-dense neighbourhoods in some countries appeared in areas dominated by social/public housing. In this way some neighbourhoods in the cities have obtained a large proportion of ethnic minorities and have been transformed into what we have called *multi-ethnic neighbourhoods*, in which ethnic social networks have appeared. Preferences for living in neighbourhoods with fellow countrymen could be important in influencing which tenures and dwellings ethnic minorities try to obtain and which kind of accommodation they can get access to.

Defective knowledge about how to act in the housing market and missing social contacts to decision makers

In parts of the housing market, good contacts to persons or institutions are decisive for access to dwellings. This especially concerns private landlords. It is important to have relevant knowledge of the possibilities and rules on the housing market, which also often requires good language skills or good access to advisers. Some studies explain the housing situation of ethnic minorities primarily via their lack of cognitive, political and social resources (Van Kempen 2003). It is particularly these non-economic resources that newly arrived ethnic minorities often lack. Several studies have shown that ethnic minorities often have difficulties in understanding rules and procedures that are decisive for getting access to housing. Bowes et al. (1997) reported from a study among Pakistani immigrants that many had virtually no knowledge at all of the allocation system used in British council housing and had fear that the allocation system was not geared towards their specific needs, especially fear of being assigned to neighbourhoods where they risked harassment from natives. Søholt (2014) uncovered that especially newly arrived refugees had severe problems in understanding how the Norwegian society and the housing market worked, which made it very difficult to get access to housing. In the Nordic comparative study by Skovgaard Nielsen et al. (2015), several interviewed Somalis in Oslo and Helsinki did not know how to balance and navigate the different access conditions to private and public rentals. They interpreted their housing situation as beyond their control. The biggest problem seemed to be a lack of transparency in allocation procedures. An exception from these negative perceptions of daunting allocation systems was reported from Denmark, where Skovgaard Nielsen (2017) found that Turks and Somalis perceived the rules for access to social housing as being easy to understand and use. In Denmark there has been a system with a simple waiting list open for all. The system has, however, been somewhat undermined in recent years by the use of alternative waiting lists in multi-ethnic neighbourhoods only for employed people.

Discrimination

Several studies (Aalbers 2002; Özüekren and Van Kempen 2003; Bowes et al. 1997; Phillips 2006; CECODHAS 2007; Andersson 1998; Søholt and Astrup 2009) point to discriminatory practices in the housing market that disadvantage ethnic minorities. In particular, social/public and especially private landlords to a certain extent exclude ethnic minorities from their housing. Moreover, discrimination from banks and real estate agents might be one of the reasons for lower homeownership rates. While strong evidence of the last kind of discrimination has been found in the US (Apgar and Calder 2005), evidence in Europe on this is more modest. In the EU-MIDIS (2009) study covering the whole EU in 2008, it was concluded that 'Discrimination when trying to open a bank account or get a loan from a bank emerged as the least problematic of the nine areas surveyed in EU-MIDIS – however, one explanation for this could be that those minorities who come into contact with banks are probably the least disadvantaged within their communities'. On average, 7 per cent of Romas, 6 per cent of North Africans and less than 5 per cent of other general groups that were surveyed identified discrimination in relation to opening a bank account or trying to obtain a loan. Stefan et al. (2018) examined how 1,218 banks in seven countries responded to applications for loans from people with respectively European and Arabic names and found pronounced discrimination. A study in Britain (Kara and Molyneux 2017) concluded that, on average, the probability of obtaining a mortgage was similar for white and non-white households. However, they found that black households with low incomes were less likely to have mortgages compared to white households with similar characteristics. Asian households, in contrast, did not seem to have a lower probability of having a mortgage.

The EUMC:67 report (EU-MIDIS 2009) concluded that there is substantial evidence of persistent discrimination against minorities in the housing field, but in some countries the number of reported cases is very low. The EUMC report distinguished between *direct discrimination, indirect discrimination* and *structural discrimination.* Direct discrimination in the report was defined as occurring where a minority person or household experiences disadvantaged treatment more frequently than native citizens. The studies across countries found a wide range of forms of direct discrimination by different actors in the housing market. Evidence showed that 'hidden' maximum quotas, for example for house hunters without employment, is in praxis used to limit and restrict the concentration of immigrant families on housing estates. This is used in countries like Germany, Finland and Denmark. Indirect discrimination means that regular or normal housing practices, requirements and conditions adversely result in the exclusion of minority households. It is stated that structural discrimination takes many various expressions and is widespread. There are, according to the EUMC:67 report, many forms of generally imposed criteria used in the allocation of social housing, which have been shown to result in limited access for, and exclusion of, migrants and ethnic minorities in the housing market.

The data showed in Chapter 8 on discrimination of certain ethnic groups across European countries reveals that discrimination varies a lot across ethnic groups and countries. On average North Africans seem to be the most discriminated against group, followed by Sub Saharan Africans and Turks, but the differences can to some extent be explained by differences across the countries where they live.

The extent and character of housing supply accessible for ethnic minorities

The options for ethnic minorities in the housing market and the difficulties they meet in their efforts to get decent housing are very much dependent on the supply of housing in different housing tenures and how housing policy regulates access and housing costs. As is shown in Chapter 8, the supply and character of rental and social housing diverges greatly across European countries. The supply of low-cost rental housing, especially social housing, is very important for ethnic minorities, but there are also some other properties of the housing market that have importance.

Exacerbating the disadvantage caused by low incomes, ethnic minorities can have particular difficulties in accessing decent housing if the housing market is un-transparent and difficult to understand. It is likely to be more difficult for ethnic minorities with a limited knowledge of the host society to act on un-transparent housing markets and find good solutions to their housing needs (Søholt 2007; Søholt and Astrup 2009).

The supply of social housing and rules for access vary greatly across countries. Sometimes social housing is open for all house hunters and sometimes access is restricted to families with low incomes or specific needs. There is usually a connection between extent of supply and rules for access where countries with small social housing sectors also have stronger restrictions on access. As the next section shows, regarding the experiences from four Nordic countries, large social housing sectors with free access are the most favourable for ethnic minorities. In some cases, however, restricted access can be an advantage for them if the supply of housing is sufficient to meet their demand, as the example of Finland shows. It can also be an advantage for ethnic minorities if local authorities can dispose of some of the vacant dwellings in social housing. In some countries this option has been used to house vulnerable ethnic minorities and especially newly arrived refugees (CLIP 2007).

The extent to which discrimination occurs depends on the way housing tenures are regulated and supported through housing policy. Sometimes there has been direct or indirect discrimination incorporated in rules for access to social housing or for housing benefits. Indirect discrimination has occurred when using criteria to reduce the proportion of ethnic minorities and unemployed in deprived and multi-ethnic neighbourhoods, a so-called mixing policy (Bricocoli and Cucca 2016; Holmqvist and Bergsten 2009). In most cases social criteria are used, for example in Antwerp, Budapest, Copenhagen and Dublin (CLIP 2007). The CLIP study showed that these policies in some cases had a discriminatory character with fixed quotas for the proportion of ethnic minorities in certain housing estates (e.g. in Stuttgart 80 per cent of the residents should be from EU countries). Another type of discriminatory rule is using the duration of residence in the city as a criterion (CLIP 2007), or the character and duration of residence permits (as in Britain reported by Rotter and Latorre 2009).

Moreover, if access to housing is very dependent on decisions made by local administrators of housing and subject to local discretion, there is greater scope for discrimination than if there are general, strict and non-discriminating rules for allocation of vacant dwellings. This is mostly the case for private renting. Moreover, the pressure of demand for the housing in question is important; if demand greatly exceeds supply there will be queues, which will tend to generate conditions more conducive to discrimination. This is illustrated in the results from the Nordic study described subsequently. One manifestation of discrimination in the private rental market is that ethnic minorities may have to pay more than the native population, which has been found in Norway (Røed Larsen and Sommervoll 2011). Surplus demand can either be a result of price and rent regulations, which keep rents and prices below market levels, or it could be due to a housing shortage in general or that the supply of publicly supported housing for low-income groups is too low.

Housing markets diverge greatly across European countries and it is therefore a complex and comprehensive task to uncover all the different housing systems used and how they affect ethnic minorities. In the following section we will go through a case study concerning four Nordic countries based on Skifter Andersen et al. (2013) and Skifter Andersen (2012) to illustrate some of the aspects of the special importance of housing policy for ethnic minorities.

A comparative study on the special importance of housing policies for the housing situation of ethnic minorities in four Nordic countries

The purpose of the study (Skifter Andersen et al. 2013) was to examine the reasons why ethnic minorities' housing position differs so much across four countries, Denmark, Finland, Norway and Sweden, and what can be learned from this comparison about the effects of housing policy on ethnic minorities' housing choices and options. These countries all belong to what is called the Social Democratic welfare model, distinguished from other welfare state models by their tendency to have greater levels of equality. However, these countries have quite different housing policies and housing markets (Bengtsson et al. 2006; Skifter Andersen 2012). All four have experienced substantial immigration for the last 25 years, which has changed the composition of the population. All the countries have developed policies for integration of new residents alongside with and interwoven with their universal welfare policies. Despite that, the majority of the ethnic minorities in all four countries find themselves in different housing positions than the ethnic majority population. Moreover, ethnic minorities' housing position differs a great deal across the countries, both in absolute terms and compared with that of the native population.

It has been demonstrated in several studies (see Chapter 2) that different ethnic groups can have varying housing preferences and perform differently on the housing market. There are differences especially between immigrants coming from Western countries and those coming from Non-Western countries. When we in the following look at explanations for differences between the four Nordic countries, it is therefore important to be aware of the differences in the immigration to the countries. As described in Andersson et al. (2010), there are some important distinctions in the nature of the immigration to the four countries. The extent of immigration varies somewhat. Finland has received fewer immigrants than the other countries, while the share of ethnic minorities in Sweden is about 40 per cent higher than in Denmark and Norway.

Moreover, the ethnic composition of ethnic minorities varies across the countries. Most of the immigrants to Finland come from Eastern Europe, especially from Russia. Finland has relatively few immigrants from countries outside Europe compared with the other three countries. Norway, Denmark and Sweden also have some immigrants from Eastern Europe.

Housing policies in the four Nordic countries

Danish and Swedish housing policies are characterised by being more general and universalistic. In Norway and Finland there are, to a greater extent, special policies for low-income groups regulated by means tests. Based on Andersson et al. (2010), Skifter Andersen et al. (2013) and Skifter Andersen (2012), Table 6.1 roughly illustrates the differences between the housing systems in the four countries.

As shown in the following, Sweden and Finland have smaller income differences between residents in renting and ownership and thus have more unitary housing systems, but the rental market in Finland is more dual because social housing has been a restricted sector for low-income groups. Norway and Denmark have a stronger social separation between renting and owning and thus more dual systems. The rental sector in Denmark is more unitary with a large social housing sector, which is not socially restricted, but rent control in private renting results in surplus demand and queues for this form of tenure. So in a way, the rental market is dual for ethnic minorities. In Norway, the rental sector, and especially the social housing sector, is quite small, which is the reason why it is more segmented. Large cooperative sectors both in Sweden and Norway form a bridge between renting and owning, which to some extent makes the housing systems more unitary.

Table 6.1 A rough characteristic of the housing systems in four Nordic countries

	Rental markets	
Housing system	*More unitary*	*More dual*
More unitary	Sweden	Finland
More dual	Denmark	Norway

Sources: Skifter Andersen et al. 2013 and Skifter Andersen 2012

Note: For ethnic minorities, the Danish rental market is more dual because of difficult access to private renting.

Political objectives concerning the desirability of different kinds of housing tenures have differed. Sweden is the only country that has had an explicit objective to give equal status to all tenures. In Norway, it has been an explicit political notion that homeownership is the most desirable kind of housing for all. Rental housing is regarded as a temporary stage in the housing career. Denmark and Finland have not formulated political objectives concerning tenures, but in practice their policies have been most favourable for owner-occupation (Skifter Andersen 2012).

The provision of a social/public housing sector has been carried out somewhat differently in the countries. Twenty years ago, Sweden had the largest social/public housing sector, but conversion into cooperatives has reduced the sector to 14 per cent of the housing stock. Finland has about the same amount of social/public housing, while Denmark has the largest sector (21 per cent). In Norway, the social/public housing sector is very small, only about 5 per cent of the stock.

Instead of social/public housing, Norway has historically focused on cooperative housing as housing for middle and low-income groups. Today prices for cooperative housing in the metropolitan region have increased to the same level as homeownership. The main public support is to supply low-income households with subsidised state loans and housing allowances in order to be able to buy and keep a dwelling, regardless of the nature of ownership. The cooperative sector is of very little importance in Finland and is also only a small sector in Denmark, except for the City of Copenhagen. Sweden has the largest cooperative sector. It can to some extent be seen as a substitution for owner-occupied flats, which to a very limited extent exist in Sweden, and the average incomes in the sector are above average. The history of cooperatives in Norway and Sweden is somewhat similar, and in both countries the sector is dominated by large housing associations. Cooperatives in Denmark are quite different. They have mostly been converted from private renting and they have been subject to a strict price control, which has to some extent been lifted in recent years. Overall, the rental sector is largest in Denmark, with about 40 per cent, followed by Sweden with 30 per cent, Finland with 30 per cent and Norway with 23 per cent.

In all of these countries private renting has been declining over time in both absolute and relative size. In Denmark and Norway, it is still about one fifth of the stock, while in Sweden it now accounts for 15 per cent of dwellings and in Finland 14 per cent. There are, however, significant differences between the countries concerning the conditions for private renting. Denmark still has a strict rent control in a major part of the stock; Sweden has a modest control; while rents in Norway and Finland are determined by the market. In Denmark and Sweden, rent regulation to some extent results in rents below market rents. This is especially the case in Denmark, where rents are very differentiated with at the same time very low rents in part of the market and very high rents in others.

The size of the owner-occupied sector differs somewhat as a more or less direct effect of housing policies. Norway and Finland have the largest sectors (62 and

59 per cent, respectively). In Denmark and Sweden, owner-occupied housing accounts for somewhat more than half of the stock. However, if cooperatives are viewed as a kind of owner-occupation, Norway has the largest share with 76 per cent, followed by Sweden with 70 per cent.

Income segmentation of the housing markets

Table 6.2 compares the proportion of homeowners in different income quartiles across the four countries. An index for income segmentation for each country has been calculated as shown at the bottom of each table.

The table shows that despite the declared objective in Norway to promote home-ownership, a much lower proportion of the households in the lowest household income quartile are homeowners compared with the other countries. In the highest income quartile, the proportion is much higher. The calculated index of segmentation is much higher in Norway than in the other three countries, depicting strong income segmentation and a more dual housing market. Skifter Andersen et al. (2013) also shows that the personal incomes in rented housing are relatively lower in Norway than in the other countries. The housing market in Denmark is not quite as segmented as the Norwegian one, but more segmented than in Sweden and Finland.

Access and discrimination of ethnic minorities in the housing market

The access to social/public housing in the Nordic countries is determined by the general rules for allocation of vacant dwellings and the extent to which local authorities have the right of disposal of dwellings to allocate to low-income groups. In all countries, local authorities have an obligation to provide housing for new refugees, which most often results in settlement in social/public housing.

Table 6.2 Homeownership in different income quartiles, 2005/07, and an index for income segmentation of housing markets

	Denmark	*Finland*	*Norway*	*Sweden*
Household income quartile		Homeownership, %		
1	50	53	31	51
2	52	65	54	60
3	66	71	74	62
4	82	74	88	71
All	62	67	62	59
Index for income segmentation	**29**	**18**	**48**	**14**

Sources: First European Quality of Life Survey: Social dimensions of housing www.eurofound.europa.eu/pubdocs/2005/94/en/1/ef0594en.pdf and Norway: Levekårsundersøkelsen 2007

Index of segmentation = sum(abs(proportion of homeowners in quartile – proportion of homeowners for the whole population))/proportion of homeowners for the whole population*100

In Denmark and Sweden, access to social/public housing is in principle based on an open allocation system with waiting lists. In recent years, however, the access in Denmark has become somewhat more difficult for ethnic minorities without employment due to new rules of allocation that have been introduced in estates with many ethnic minorities. This is part of a policy that tries to disperse ethnic minorities across neighbourhoods and reduce them in concentration areas. In Finland as in Norway, access is based on urgent housing needs, which favours the most vulnerable minorities, but not the more successful.

In Denmark, rent control in the private rented sector produces queues that do not favour ethnic minorities and tend to reduce their possibilities of accessing the sector, because ethnic minorities seldom know landlords personally and because some landlords do not want to rent to ethnic minorities (Skifter Andersen 2008). In Norway, the small rental sector increases the competition between house hunters who want to rent, which provides a fertile environment for discrimination. Moreover, the market is dominated by small landlords letting one or a few dwellings, who tend to avoid ethnic minorities as tenants (Nordvik 1996; Søholt and Astrup 2009). Finland has a more professionalised private rental sector, which puts more weight on allocation by market mechanisms and should facilitate ethnic minorities' access if they can pay the rent. In Sweden, it is permissible for municipalities to establish a central housing allocation service for private renting, which is open to ethnic minorities, but the landlords are not obliged to use it, so here personal contacts with landlords are also important.

Several qualitative studies on the housing problems of ethnic minorities in the capitals of the four countries have been conducted, and these illuminate difficulties in getting access to housing and discrimination (Dhalman and Vilkama 2009; Søholt 2014; Skovgaard Nielsen 2017; Skovgaard Nielsen et al. 2015). The comparative study of Somalis conducted by Skovgaard Nielsen et al. (2015) found significant differences between the capitals concerning perceived discrimination. In Copenhagen complaints regarding discrimination were very limited. However, the experiences of ethnic minorities were mostly limited to social housing, where the allocation rules are specified and combined with a system of transparent waiting lists. It appeared that ethnic minorities mostly did not perceive private renting as a possible option (Skovgaard Nielsen 2017), probably because the social housing sector was quite easy to access. In the other three capitals, on the contrary, believed discrimination and racism were reported. Some landlords in Oslo refused Somali applicants by saying that they did not let to ethnic minorities and especially not to Somalis. This is in line with earlier studies (Blom and Henriksen 2008; Søholt and Astrup 2009). Others searching for a dwelling in the private market had encountered significant difficulties, which they assumed were caused by discrimination.

Cooperatives in Denmark differ greatly from cooperatives in Sweden and Norway. Cooperatives in Denmark have been subject to price control that tends to produce prices beyond the market level. Access to Danish cooperatives has been very difficult for ethnic minorities, because allocation of dwellings has been controlled by the boards of the cooperatives that often have chosen family or friends as new residents. In Norway, access to cooperatives is market based, combined with seniority as a member of the association. The applicants have to be formally accepted by the boards of the cooperatives, but there is not much room for discrimination. In Sweden, allocation is market-based. In new construction, which earlier was given subsidies, waiting lists are used with seniority as the allocation principle.

Access to owner-occupied housing is market-based in all the countries. There is not much research on discrimination against homebuyers in the Nordic countries. Avoiding-strategies from sellers, neighbours and real estate agents are, however, documented in Norway, but seem to be less frequent than discrimination in the rental market (Søholt 2007; Søholt and Astrup 2009). However, difficulties in obtaining loans or mortgages for housing have to some extent been reported (Skovgaard Nielsen et al. 2015). This depends very much on the kinds of finance and subsidies available in the various.

In Skifter Andersen et al. (2013) and Skifter Andersen (2012), an evaluation of the ease with which ethnic minorities can access different tenures in the countries was conducted. It takes into account the issues identified previously concerning supply of different tenures and of the described rules for access and practices of administration. Three factors were considered in determining these evaluations: the relative size of the tenure, which especially applies to the rental sectors; specific rules regulating access; and finally, whether conditions on the market make room for discrimination of ethnic minorities. The conclusion was that Denmark has the most differentiated housing market with the best opportunities for access to social housing. This was most difficult in Norway. In all four countries, private renting was the least accessible tenure for ethnic minorities. Access to cooperatives was most difficult in Denmark due to price regulations, the importance of network allocation and queues.

This picture was supported by the comparative qualitative study made by Skovgaard Nielsen et al. (2015), which showed that Somalis in Copenhagen and Helsinki were more content with their housing conditions than those in Stockholm and Oslo. In Copenhagen, they generally perceived their housing situation and possibilities within the housing market to be good. In Helsinki, housing conditions were also generally perceived to be good except for overcrowding and difficulties in getting a flat. The former was caused by a lack of units of appropriate size for large families and by homeless staying with friends and relatives. In the Stockholm case, half of the interviewees told of poor housing situations, usually related to overcrowding and poor maintenance of the dwelling or the living environment. In Oslo, some experienced satisfactory housing conditions, while others faced overcrowding and/or lack of maintenance. The main problems in Oslo were shortage of rentals. Difficult access to decent housing resulted often in living with acquaintances and time-limited contracts making it difficult to establish a home. Most of the interviewees in Oslo were struggling financially to pay their rent and to afford a suitable dwelling. Those in temporary shelter had to renew their contract every second week.

Comparison across countries of ethnic segmentation and ethnic minorities' housing situation

Ethnic segmentation in the housing market can be examined by comparing the distribution of ethnic minorities on tenures with the distribution of the whole population. Table 6.3 shows how ethnic minorities in the countries are distributed across tenures, and their over or under-representation in each form of tenure. Moreover, an index of ethnic segmentation for each country is shown. The index is calculated as the sum of the absolute value of the difference between the proportion of ethnic minorities in each tenure and the proportion of the whole population living in that tenure, weighed by the proportion of the whole population living in the tenure.

The table shows significant differences across the four countries regarding ethnic minorities' distribution on tenures. In particular Norway differs by having a substantially larger proportion of ethnic minorities living in owner-occupied housing. In the other three countries, homeownership is at about the same level, accounting for about 30 per cent. In all the countries, ethnic minorities are under-represented in the owner-occupied tenure, but as homeownership is not at the same level in the various countries, ethnic minorities are, compared to the whole

Table 6.3 Non-Western ethnic minorities distributed on tenures in four Nordic countries (%) plus ethnic minorities' over-representation in each tenure and a calculated index of tenure segmentation

	Denmark	*Finland*	*Norway*	*Sweden*
Ethnic minorities				
Owner-occupied	30	31	45	30
Cooperatives	6		18	23
Private renting	16	21	24	26
Social housing	45	43	13	20
Other	2	5		
Total	100	100	100	100
Over-representation (%)				
Owner-occupied	−51	−57	−27	−45
Cooperatives	−4		29	37
Private renting	18	91	26	89
Social housing	166	222	160	45
Other	113	40		
Index of ethnic segmentation	**63**	**81**	**34**	**50**

Year and sources: Denmark: 2008, Database at the Danish Building Research Institute based on data from Statistics Denmark. Finland: Statistics Finland 2010. Norway: Living conditions among immigrants in Norway 2005/2006 and living conditions among the whole population 2007, survey data Statistics Norway. Sweden: GeoSweden 2008

Note: Over representation = (% of ethnic minorities in tenure x – % of whole population in tenure x)/ % of whole population in tenure x

Index of ethnic tenure segmentation = $sum_{x=1,n}$ (Absolute (% of ethnic minorities in tenure x – whole population in tenure x)/proportion of population in tenure x*100)

population, less often homeowners in Finland than in Denmark and Sweden. This could partly be due to the fact that ethnic minorities in Finland have arrived later than in the other countries and have had less time to 'climb up the housing ladder' (Kauppinen et al. 2015).

There are also remarkable differences between the countries concerning ethnic minorities' representation in the other tenures. In Denmark and Finland, a large proportion of them live in social housing, while it is only 20 per cent in Sweden and 13 per cent in Norway. The low proportion in Norway is due to the very small social housing sector. When it comes to ethnic minorities' over-representation in social housing, Norway is at the same level as Denmark, while Finland has the largest over-representation. Ethnic minorities are also over-represented in social housing in Sweden, but at a lower level than in the other countries.

Private renting is an important sector for ethnic minorities in all the countries studied, accommodating between 15 and 25 per cent. But compared with the size of this tenure, ethnic minorities have much more often gained access to private renting in Finland and Sweden. Especially in Denmark, ethnic minorities do not very often live in private renting. A Danish study (Skifter Andersen 2010) has shown that after taking into account that their incomes are much lower than average, ethnic minorities have a 60 per cent lower probability of moving into private rented dwellings. They thus less often live in private renting in Denmark than what would be expected considering their lower incomes.

In Norway and Sweden, cooperatives are important for ethnic minorities, who are over-represented in the sector in both countries, especially in Sweden. One reason could be that there is little room for discrimination in the rules for access. Again, the Danish case greatly differs from the others, because ethnic minorities less often live in cooperatives than the whole population. Considering their incomes, the probability of ethnic minorities moving into cooperatives is even much smaller because low-income groups in Denmark much often move into cooperatives (Skifter Andersen 2010).

The calculated index of segmentation shown at the bottom of Table 6.3 indicates substantial differences between ethnic minorities' position on the housing market in the four countries. Finland is the country where the largest deviation is found between ethnic minorities' housing tenure and the rest of the population. Denmark has the second largest deviation followed by Sweden. In Norway, ethnic minorities' housing tenure is most like the national average.

The high proportion of ethnic homeowners in the Norwegian housing system has, however, had a price. As can be seen from Table 6.4, ethnic minorities in Norway, more often than in the other countries, live in overcrowded dwellings, especially compared with the average for the whole population. The table shows figures for overcrowding measured as the proportion of residents living in dwellings with less than one room per person. The relative difference between ethnic minorities and the whole population is calculated as the percentage by which overcrowding for ethnic minorities exceeds overcrowding for the whole population.

Table 6.4 Proportion of ethnic minorities and whole population living in densely populated housing* in the Nordic countries (%)

	Denmark	*Finland*	*Norway*	*Sweden*
Whole population	7	18	8	16
Ethnic minorities	28	41	45	26
Relative difference %	289	128	463	61

Sources: Denmark: Foreign born persons and their children 2008 16+ years. Database at the Danish Building Research Institute based on data from Statistics Denmark. Finland: Data on households 31.12.2008, ethnic minorities = households headed by a foreign-speaking person, Source: Statistics Finland. Norway: Source: Foreign-born persons and their children. Age 16–70 years. Statistics Norway. Level of living among ethnic minorities 2005/2006. Survey to a representative selection among 10 groups. Level of living in the whole population 2007. Single adults in one room dwellings excluded. Sweden: Foreign-born persons 16+ years. Statistic Sweden, Survey on living conditions 2007

*More than one person per room (kitchen and bath excluded).

Ethnic minorities in Norway live five to six times more often in overcrowded housing compared with the whole population. Ethnic minorities in Finland are also quite often living in overcrowded housing, but this can partly be explained by a generally high degree of overcrowding in the country. Ethnic minorities' level of overcrowding is close to the same level in Denmark and Sweden. However, because the population in Denmark in general more seldom lives in overcrowded dwellings, the difference between ethnic minorities and the whole population is much greater than in Sweden. It can therefore be concluded that ethnic minorities' housing situation, when it comes to overcrowding, differs most from that of the whole population in Norway followed in decreasing order by Denmark, Finland and Sweden.

The special importance of housing policy in the Nordic countries

The comparison of the four Nordic countries shows that differences in the position of ethnic minorities in the housing market cannot be explained by differences across the countries with respect to immigration and general income inequalities alone. It indicates that there is not a straightforward connection between income segmentation of housing markets and ethnic segmentation. On the contrary it turned out that the country with the highest income segmentation (Norway) had the lowest ethnic segmentation, and that the country with the lowest income segmentation (Finland, except for Sweden) had the highest ethnic segmentation. This indicates that housing policies have special effects for the housing situation of ethnic minorities.

In connection with the discussion of the effects of dual housing markets and dual rental markets for the distribution of ethnic minorities across tenures, these results show an unexpected pattern. It should be expected that Norway, where both a more dual housing market and a more dual rental market are found, should be most ethnically segmented, but the opposite is true. The other countries are more in accordance with expectations, but it seems that the character of rental markets is the most important factor, as Finland has higher segmentation than both Denmark and Sweden. Sweden and Denmark both have a more unitary rental market, but the lower segmentation in Sweden is in accordance with a more unitary housing market in general compared to Denmark.

The comparison shows that the main factors explaining ethnic segmentation are the size of the rental sector, the size of the social housing sector and rules for access to this sector, the character of private renting and the role of cooperative housing. The Norwegian case shows that a lack of sufficient rental housing in a country can push ethnic minorities into owner-occupation, which result in less ethnic segmentation and could be seen as desirable. But the price for this policy has been that many of them live in overcrowded accommodations. Moreover, the lack of rental housing results in a high demand pressure on private renting. This makes room for increased rents but also discrimination, which has been well documented in the Norwegian case (Røed Larsen and Sommervoll 2011; Søholt and Astrup 2009).

In the other three countries, there is more rental housing in general and more social housing, especially in Denmark, but these sectors have been differently organised. Important differences for ethnic minorities are rent and price controls in private renting and cooperatives, the size of the social/public housing sector and rules of access to social/public housing. The Danish case shows that ethnic minorities' housing options and experience of discrimination are influenced by regulation of the private rental market and a lack of transparency in housing transactions. Strong rent control, as is the case in Denmark, reduces the housing possibilities of ethnic minorities in private renting because rents below the market level create queues, which give landlords options to choose between house hunters. They tend to choose family and acquaintances as tenants, and as ethnic minorities seldom have good informal contacts with landlords, they experience difficulties in gaining access to these dwellings. Also, price control in cooperatives in Denmark has resulted in a strong under-representation of ethnic minorities in this tenure because of stronger competition between house hunters, which ethnic minorities tend to lose in absence of personal contacts to members of the cooperatives.

A large social housing sector, such as that found in Denmark, and prior to 2000 in Sweden, paves the way for good housing opportunities for ethnic minorities. Open access to social housing for all income groups increases ethnic minorities' housing options. However, the lessons from Finland are that if housing policy is concentrated on support for housing for the poor, who are housed in social housing allocated strictly on the basis of need, ethnic minorities tend to be very concentrated in this kind of housing. Also, in Denmark there is a high concentration of ethnic minorities in social housing. This can only partly be explained by a housing policy that in general has resulted in higher income segmentation in the housing market than in Sweden and Finland. More important is that ethnic minorities have less access to private renting and cooperatives. The universalistic housing policy in Sweden of giving equal status to all tenures has resulted in a modest ethnic segmentation of the housing market and in a smaller over-representation of ethnic minorities in social housing. It has also resulted in smaller differences between ethnic minorities and the native population concerning overcrowding.

Note

1 Lena Magnusson Turner, NOVA Institute, Oslo University, Norway and Susanne Søholt Norwegian Institute for Urban and Regional Research, Norway have contributed with comments and with data to the Nordic comparative study, which has been used in this chapter.

References

Aalbers M. B. (2002) The Neglected Evidence of Housing Market Discrimination in the Netherlands, *Radical Statistics Journal*, 79/80, pp. 161–169.

Andersson R. (1998) Socio-spatial Dynamics: Ethnic Divisions of Mobility and Housing in Post-Palme Sweden, *Urban Studies*, 35(3), pp. 397–428.

Andersson R. et al. (2010) *Immigration, Housing and Segregation in the Nordic Welfare States*. Department of Geosciences and Geography, University of Helsinki.

Apgar C. A. and Calder A. (2005) *The Dual Mortgage Market: The Persistence of Discrimination in Mortgage Lending*. W05–11 Joint Centre for Housing Studies Harvard University. Website: at www.jchs.harvard.edu/sites/default/files/w05-11.pdf.

Bengtsson B. Annaniassen E., Jensen L., Hannu Ruonavaara H. and Sveinsson J. R. (2006) *Varför så olika? Nordisk bostadspolitik i jämförande historisk ljus*. Malmö: Egalitè.

Blom S. and Henriksen K. (2008) *Levekår blant innvandrere i Norge 2005/2006 (Living Conditions among Immigrants in Norway 2005/2006)*. Report 5. Oslo: Statistics Norway.

Bowes A., Dar N. and Sim D. (1997) Tenure Preference and Housing Strategy: An Exploration of Pakistani Experiences, *Housing Studies*, 12(1), pp. 63–84.

Bricocoli M. and Cucca R. (2016) Social Mix and Housing Policy: Local Effects of a Misleading Rhetoric. The Case of Milan, *Urban Studies*, 53(1), pp. 77–91.

CECODHAS (2007) *Housing Europe 2007 Review of Social, Co-operative and Public Housing in the 27 EU Member States*. Brussels: CECODHAS Housing Europe.

CLIP (2007) *Housing and Integration of Migrants in Europe*. Dublin: European Foundation for the Improvement of Living and Working Conditions.

Dhalmann H. and Vilkama K. (2009) Housing Policy and the Ethnic Mix in Helsinki, Finland: Perceptions of City Officials and Somali Immigrants, *Journal of Housing and the Built Environment*, 24, pp. 423–439.

Doling J (1997) *Comparative Housing Policy*. Basingstoke: MacMillan Press.

EU-MIDIS (2009) *European Union Minorities and Discrimination Survey Main Results Report*. European Union Agency for Fundamental Rights.

EWSI Analysis (2016) *Immigrant Housing in Europe*. European Commission: European Web Site on Integration. Website: at https://ec.europa.eu/migrant-integration/.

Finney N. (2002) Ethnic Group Population Change and Interaction: A Demographic Perspective on Ethnic Geographies. Ethnicity and Integration, *Understanding Population Trends and Processes*, 3, pp. 27–45.

First European Quality of Life Survey: Social dimensions of housing. European Foundation for the Improvement of Living and Working Conditions. Retrieved from www.eurofound.europa.eu/pubdocs/2005/94/en/1/ef0594en.pdf

Ghekiere L. (2007) *Le développement du logement social dans l'Union européenne*. Caisse Nationale des Allocations Familiales, Paris. Website: at www.caf.fr/sites/default/files/cnaf/Documents/Dser/PSF/094/RP94-LGhekiere.pdf

Holmqvist E. and Bergsten Z. (2009) Swedish Social Mix Policy: A General Policy without an Explicit Ethnic Focus, *Journal of Housing and the Built Environment*, 24(4), pp. 477–490.
Johnston R., Forrest J. and Poulsen M. (2002) Are there Ethnic Enclaves in English Cities? *Urban Studies*, 39, pp. 591–618.
Kara A. and Molyneux P. (2017) Household Access to Mortgages in the UK, *Journal of Financial Services Research*, 52(3), pp. 253–275.
Kauppinen T. M., Skifter Andersen H. and Hedman L. (2015) Determinants of Immigrants' Entry to Homeownership in Three Nordic Capital City Regions, *Geografiska Annaler: Series B, Human Geography*, 97(4), pp. 1–20.
Kemeny J. (1995) *From Public Housing to the Social Market: Rental Policy Strategies in Comparative Perspective*. London: Routledge.
Kemeny J. and Lowe S. (1998) Schools of Comparative Housing Research: From Convergence to Divergence, *Housing Studies*, 13(2), pp. 161–176.
Levekårsundersøkelsen (2007). Statistisk Sentralbyrå, Oslo.
Lindberg G. and Lindèn A.-L. (1989) *Social segmentation på det svenske bostadsmarknaden* (Social Segmentation on the Swedish Housing Market). Sociologiske Institutionen, Lunds Universitet, Lund.
Musterd S. (2005) Social and Ethnic Segregation in Europe: Levels, Causes and Effects, *Journal of Urban Affairs*, 27(3), pp. 331–348.
Nordvik V. (1996) Mot et mer ustabilt leiemarked. *Samfunnsspeilet*, 4, pp. 15–18.
Özüekren A. S. and Van Kempen R. (2003) Special Issue Editors' Introduction: Dynamics and Diversity: Housing Careers and Segregation of Minority Ethnic Groups, *Housing, Theory and Society*, 20(4), pp. 162–171.
Phillips D. (2006) Parallel Lives? Challenging Discourses of British Muslim Self-segregation, *Environment and Planning D: Society and Space*, 24, pp. 25–40.
Røed Larsen E. and Sommervoll D. E. (2011) Prisdannelsen i det norske leiemarkedet: en teoretisk og empirisk analyse av hovedmekanismer generelt og utsatte grupper spesielt. In NOU 2011:15 *Rom for alle. En sosial boligpolitikk for framtiden*. Oslo: Ministry of Local Government and Regional Development.
Rosenfeld O. (2015) *Social Housing in the UNECE Region. Models, Trends and Challenges*. Geneva: United Nations Economic Commission for Europe.
Rothenburg J., Galster G. C., Butler R. V. and Pitkin J. R. (1991) *The Maze of Urban Housing Markets: Theory, Evidence and Policy*. University of Chicago Press.
Rotter J. and Latorre M. (2009) *Social Housing Allocation and Immigrant Communities*. Manchester: Equality and Human Rights Commission.
Skifter Andersen H. (2012) *The Importance of Housing Policies and Housing Markets for the Housing of Immigrants in the Nordic Countries*. Horsholm: Danish Building Research Institute at Aalborg University.
Skifter Andersen H. (2011) Motives for Tenure Choice during the Life Cycle, *Housing, Theory and Society*, 28(2), pp. 183–207.
Skifter Andersen H. (2010) Spatial Assimilation in Denmark. Why do Immigrants Move to and from Multi-ethnic Neighbourhoods? *Housing Studies*, 25(3), pp. 281–300.
Skifter Andersen H. (2008) *Privat boligudlejning. Motiver, strategier og økonomi*. Horsholm: Statens Byggeforskningsinstitut.
Skifter Andersen H., Magnusson Turner L. and Søholt S. (2013) The Special Importance of Housing Policy for the Housing Situation of Ethnic Minorities, *International Journal of Housing Policy*, 13(1), pp. 20–44.
Skifter Andersen H. and Munk A. (1994) The Welfare State Housing Policy versus the Social Market Economy - A Comparison of Denmark and Germany, *Scandinavian Housing and Planning Research*, 11(3), pp. 1–25.

Skovgaard Nielsen R. (2017) The Potentials of a Strong Social Housing Sector: The Case of Turks and Somalis in the Copenhagen Housing Market, *Housing, Theory and Society*, 34(4), pp. 458–476.

Skovgård Nielsen R., Holmquist E., Dhalmann H. and Søholt S. (2015) Comparing Somalis' Perceived Possibilities in the Nordic Capital Housing Markets: Local Context and Cultural Background, *Housing Studies*, 30(3), pp. 433–452.

Søholt S. (2014) Pathways to Integration: Cross-cultural Adaptations to the Housing Market in Oslo, *Journal of Ethnic and Migration Studies*, 40(10), pp. 1637–1656.

Søholt S. (2007) *Gjennom nåløyet – en sammenligning av tilpasninger til boligmarkedet blant hushold av pakistansk, tamilsk og somalisk bakgrunn, Oslo 1970–2003 (Through the Eye of a Needle – A Comparison of Adaptions to the Housing Market among Households with Pakistani, Tamil and Somali Background, Oslo 1970–2003)*. Thesis. Oslo: Institute for Political Science, Faculty of Social Sciences, University of Oslo.

Søholt S. and Astrup K. (2009) *Etniske minoriteter og forskjellsbehandling i leiemarkedet (Ethnic Minorities and Unequal Treatment in the Rental Market)*. NIBR-report: 2. Oslo: Norwegian Institute for Urban and Regional Research.

Stefan M., Holzmeister F., Müllauer A. and Kirchler M. (2018) *Ethnical discrimination in Europe: Field evidence from the finance industry* PLoS ONE 13(1): e0191959. https://doi.org/10.1371/journal.pone.0191959

Van Kempen R. (2003) *Segregation and Housing Conditions of Immigrants in Western European Cities*. Eurex Lecture 7 – March 13th 2003. Website: at www.shiva.uniurb.it/eurex/syllabus/lecture7/Lecture7-VanKempen.pdf.

7 The importance of urban policies and the spatial distribution of housing[1]

In this chapter we will discuss the importance of urban policies, and the spatial structure of housing markets in cities, for ethnic segregation, based on the relatively sparse literature on this subject. An advanced study in four Nordic capitals, which is referred in the last part of the chapter, has showed that an uneven spatial distribution of different housing tenures has quite a strong influence on ethnic segregation.

Spatial inequality as a cause of segregation

The literature on causes of spatial segregation has mostly focused on the effects of the development of social and cultural differentiation of the population in cities. Spatial segregation has mainly been understood as a direct consequence of how the population has been divided into separate social and cultural groups competing for attractive neighbourhoods, resulting in a spatial sorting of groups into different parts of the cities, separated from each other. In the dominant theories from the last part of the 20th century a polarised residential structure was seen as a direct result of growth and decline in global cities, where social polarisation was conceived as a product of economic restructuring (Sassen 1994; Castells 1989). There was little focus in the segregation literature on the importance of the physical, social and economic spatial structures found in different cities (South et al. 2011), except for research into a few subjects like the effects of gentrification and so-called *social mixing* policies (Musterd and Andersson 2005). Some authors (Wessel 2000; Arbaci 2007; Musterd and Ostendorf 1998; Van Kempen and Murie 2009), however, have argued for more differentiated and context-bound explanations of segregation, where spatial characteristics of cities are included as important factors. One of these characteristics is the spatial housing structure, especially the segmentation of housing markets and the spatial separation of different housing tenures.

The socio-spatial dialectic of cities

There has also been little focus on the dynamic changes that over time occur in cities creating new spatial structures and new kinds of segregation. The development of cities has occurred by an interaction between economic, social and physical changes. It has been called the socio-spatial dialectic (Soja 1980), a continuous two-way process in which people create and modify urban spaces while at the same time being conditioned in various ways by these changes. Neighbourhoods and housing, in which people prefer to live and work, or have access to, have changed continuously in pace with increased wealth and changing tastes. As a result, the social and physical structures of cities have undergone pronounced changes over the years. The growth of every city can be seen as a 'twin process of outward extension and internal reorganisation' (Knox 1995). While cities have been extended by new estates on the suburban fringe, existing urban areas have been subject to functional and physical changes; they have sometimes been converted to accommodate new uses. In some cases, existing physical structures have become obsolete and have deteriorated; in other cases they have been upgraded to meet new demands from more well-to-do residents. However, while the population composition of neighbourhoods can change over shorter periods of time – as a result of residential mobility or demographic processes (Van Ham et al. 2013) – changes in the physical fabric and the overall socio-economic status of neighbourhoods are much slower processes (Zwiers et al. 2015).

The distribution of people in space is created by interaction between social and cultural differentiation among residents and *spatial differentiation* of cities (Skifter Andersen 2003). This spatial differentiation is a product of the social, physical and functional structure of the city, a structure that is continuously changing due to economic investments and disinvestments as people and functions are redistributed in space. Cities become divided into identifiable areas that can be relatively homogeneous, but exhibit distinctive characteristics that are very different from other neighbourhoods. Such areas differ not only by virtue of different physical qualities, such as housing stock, architecture, natural attractions and access to transport facilities and service supply, but also by less tangible qualities such as social and cultural image, status and security. Preferences for living in various neighbourhoods can vary across households with different needs and life-styles, but there will always be some common values among people that result in some neighbourhoods being more appreciated by the majority, while other neighbourhoods are seen as very unattractive. The greater differences found between the attractiveness of neighbourhoods in a city, the larger the *spatial inequality* will be.

The main point I would like to stress is that segregation is not a simple consequence of social and ethnic inequality, but is also affected by spatial differentiation leading to spatial inequality. Segregation, therefore, is greatly influenced by the development of spatial differentiation in cities, and this is perhaps more important than the development in social inequality and social exclusion. Segregation and increasing spatial inequality are mutually self-perpetuating processes because the social status and cultural identity of individual urban areas are determined by the composition of residents. Spatial differentiation leads to segregation, while segregation creates spatial differences and inequality (Skifter Andersen 2003).

The historic development in ethnic minorities' settlement pattern in cities

As described in Chapter 3, the so-called Chicago School (Park 1925; Park and Burgess 1969) in the beginning of the 20th century proposed a model for how new immigrants to the US settled in the city over time. The Chicago School's ecological model was formulated at a time when immigration was dominated by low-wage European labour. Newly arrived immigrants were poor and were expected to settle in the worst slums in the centre of the cities where they could find cheap, but low quality, housing. In pace with their economic and social integration immigrants were expected to move out from the slums to neighbourhoods with better housing and neighbourhood amenities (e.g. good schools, clean streets). Especially since the 1950s such neighbourhoods in the US have been largely suburban (Alba and Logan 1992). More recent research has found that increasingly new immigrants to the US move directly into neighbourhoods located in suburbs and settle more dispersed in the city from the beginning (Alba et al. 1999; Clark and Patel 2005; Wright et al. 2005).

Even if European cities have differed from American cities (Van Kempen and Murie 2009), the same development has occurred. Earlier immigrants, mostly coming as labour immigrants or coming from colonies, tended to settle in cheap low quality housing in the central cities (Murie og Musterd 1996). In pace with urban renewal that changed the housing stock in central cities into higher quality and more expensive housing, ethnic minorities to varying degrees were displaced to either more dispersed informal and low quality housing or to social housing in the suburbs (Hamnett and Butler 2010; Malheiros 2002). In many countries, though, central city housing still is important for ethnic minorities. This can be found in cities such as Berlin, Amsterdam and Brussels (Musterd and Ostendorf 1998). According to Arbaci (2007), cities characterised by Liberal welfare regimes (see the definition of welfare types and countries in Chapter 8) have high segregation that especially is located in city centre areas. In cities characterised by a Corporatist welfare regime, the areas of ethnic over-representation are located either in the inner city or in the inner suburbs. Cities within the Social Democratic cluster have the highest concentrations outside the city centre. Although the patterns of ethnic spatial segregation in Southern European cities are not as consolidated as in Northern European cities, the position of Southern European cities is close to the Corporatist welfare cluster.

The settlement pattern of ethnic minorities has to some extent been determined by formation of ethnic enclaves, neighbourhoods dominated by ethnic minorities, where a tight ethnic network has been established that has been attractive for immigrants (see Chapter 3). There is a mutual relationship between the character of ethnic concentration and preferences for ethnic enclaves. The existence and intensity of ethnic enclaves depends on the size and composition of ethnic groups and the extent and character of ethnic segregation in countries and cities. Preferences for ethnic enclaves depend on to what extent local neighbourhoods with a sufficient proportion of one's own ethnic group, or other groups with a similar culture, are available, which is decisive for the creation of attractive enclave properties. It is thus important for the existence of enclaves that there is a sufficient number of ethnic minorities coming from the same country or with the same culture. It is also shown (Chapter 3) that such preferences are stronger for some ethnic groups than for others. Moreover, there is evidence that ethnic groups with a lower social status in the country where they reside have stronger preferences for living separated from the native majority and other groups in accordance with the American hypotheses on the importance of racial prejudices cited in Chapter 3.

Differences between countries concerning the social status of ethnic minorities could thus have an important influence on the extent of ethnic enclaves.

Formation of enclaves is, however, also dependent on the conditions in the housing market. Creation of enclaves is easier in markets dominated by home-ownership that gives a freer choice to settle compared to renting, where landlords and social housing companies regulate access. In free markets a dominance of ethnic minorities could lead to lower prices on homes in a neighbourhood because the demand from the native majority could be lower (see Chapter 5), which makes it easier for minorities with lower income to get access.

As described in Chapter 8, the degree of segregation and the character of ethnic neighbourhoods vary greatly across European countries. In most cases ethnic neighbourhoods consist of several ethnic groups. Neighbourhoods dominated by a single ethnic group have a stronger tendency to develop true ethnic enclaves, as defined previously. It is, however, also possible that mixed neighbourhoods dominated by ethnic minorities could develop some of the same advantages for ethnic minorities. In many countries, neighbourhoods dominated by ethnic minorities, so-called multi-ethnic neighbourhoods, are found in social housing in the suburbs.

The importance of the spatial distribution of housing in cities

Different kinds of housing are to a different extent attractive and accessible for different social groups. The spatial distribution of housing is thus an important factor in creating spatial inequality. It could be argued that this spatial distribution is just a mirror of housing demand from different groups, whose preferences for location create the spatial distribution of housing tenures. In this view the housing market does not influence segregation, but is a result of segregation. This can, however, be questioned, as there is a sluggishness in the change of housing structures of cities over time. Cities' housing structures have been created over a long time, whereas population characteristics and housing demand over the course of time have changed faster. The housing market structure has to some extent been created by past housing demand and might not be in accordance with actual demand. This especially concerns the central cities with old and sometimes dilapidated and obsolete housing. In many countries housing markets might, moreover, be highly regulated by national housing policies and urban planning, which interfere with the natural processes of market segregation (Atkinson and Kintrea 2000; Galster 2007; Briggs 2005; Musterd and Andersson 2005).

Some American literature (South and Crowder 1997; South et al. 2011) holds that the probability that different groups gain access to a particular type of neighbourhood is dependent on the quantity and quality of destination opportunities on local housing markets. They called this reason for ethnic segregation *the housing availability model* (South and Crowder 1997). As early as a couple of decades ago, some European authors drew attention to the matter that access to housing is important in understanding patterns of segregation (Van Kempen 1994).

As described in Chapter 6, the comparative literature on housing policy distinguishes between 'dual' and 'unitary' housing systems (Allen 2004; Balchin 1996;

Kemeny 1995). In dual systems, the housing market is socially highly divided between rented housing and homeownership, and some tenure forms are dominated by specific social groups while other groups have difficulties in getting access to the tenure. Another division is made between unitary and dual rental systems (Kemeny 1995). In the unitary rental systems, social supported housing is competing on even terms with private renting, while in dual systems public housing is a restricted sector reserved for low-income groups.

A key concept in understanding the role of housing is *segmentation* of housing markets (see Chapter 6). Housing supply is divided into different housing tenures, which have different qualities and conditions for access. In a highly socially or ethnically segmented housing market, conditions for access to different tenures differ greatly for different social or ethnic groups, resulting in an uneven distribution of different groups across housing tenures. This uneven distribution has also been called 'socio-tenurial polarisation' (Murie and Musterd 1996) and 'socio-tenure differentiation' (Arbaci 2007). Housing market segmentation depends on to what extent housing policies create even or uneven opportunities and economic incentives in different tenures, as described in Chapter 6.

The extent to which housing policy and housing market segmentation affects segregation depends on the way different kinds of housing are spatially distributed; if different tenures are concentrated in different neighbourhoods or if they are mixed. This depends in turn on politics, institutions and markets that shape and change the urban structure. Earlier research (e.g. Murie and Musterd 1996; Arbaci 2007) focused on the importance of the size and character of the social housing sector in the cities as a major explanation of differences in segregation. Arbaci (2007) argued that the relatively high level of ethnic segregation among some countries, in spite of having more uniform and less socially segmented housing markets, could be ascribed to the existence of large spatial concentrations of social housing with many immigrants. It is assumed that a marginalised social sector with high concentrations of low-income groups also will result in high ethnic segregation, but this depends extensively on the size of the sector and to what extent it is concentrated on large estates in certain parts of the cities.

It should be expected that dual housing markets with strong income segmentation also have a higher degree of ethnic segmentation, because ethnic minorities tend to have lower incomes. However, Chapter 6 and Chapter 8 demonstrate that there is not a systematic connection between income segmentation and ethnic segmentation of housing markets. One of the reasons for this is that immigrants have special conditions and behaviour in the housing markets that influence their housing options and choices, as discussed in other chapters, and these are influenced in particular ways by housing policy.

In a comparison between Britain and the Netherlands, Murie and Musterd (1996) identified changes in the size and operation of the social rented sector as very important for the development of segregation in Dutch and British cities. Kesteloot and Cortie (1998) compared the settlement patterns of Turks and Moroccans in respectively Brussels and Amsterdam and concluded that housing market variables explained many of the differences between the two cities. Based on a comparison of welfare systems, housing policies and ethnic segregation in cities in eight European countries, Arbaci (2007) concluded that differences in housing policy and the resulting housing markets in the countries, together with planning systems, have played a very important role in explaining differences in ethnic segregation across countries.

The importance of urban development and planning for the spatial structure of urban housing markets

As discussed in the beginning of this chapter, it can be argued that the spatial structure of housing markets in cities is not just a mirror of housing demand from different social and ethnic groups nor simply a result of segregation. The structure of cities has been created over a long time mirroring earlier needs and demands for use of space. Some changes have occurred driven by market forces, but often they have not been able to make cities adapt in a sufficient way to meet new preferences and demands for housing and residential qualities. In most cities some neighbourhoods have developed into slum areas and deprived neighbourhoods (Skifter Andersen 2003).

Sometimes urban renewal has been driven by market forces resulting in phenomena like gentrification, where older neighbourhoods, which earlier were run down and inhabited by low-income groups, were converted into high quality areas inhabited by people with higher incomes (Nelson 1988). In other cases renewal has been a result of public supported programmes, sometimes guided by social objectives of retaining a more socially mixed neighbourhood with more diverse housing (Skifter Andersen 1998; Skifter Andersen and Leather 1999). In many cases, however, central cities have been left with old housing of a low quality and much poor so-called informal housing, which has been available for people with low incomes and often immigrants.

Also, the development of new housing estates on the urban fringe, or on converted land, has to some extent been influenced by public planning and land control. Urban planning in cities might have had other objectives for urban development than what would be a result of market forces. In this way there sometimes has been created a more mixed spatial structure of housing markets than a pure market-governed structure. It is not certain that this planning always has led to more mixed neighbourhoods. For example, functionalist planning principles, especially as performed in the 1960s and 70s, have sometimes resulted in the creation of large urban areas with uniform housing, often social housing (Wassenberg 2013).

The role of urban renewal in European cities

Urban renewal after the war was directed towards problems of decay and obsolete housing in the older and central parts of cities. In the first period, until the start of the 1970s, slum clearance involving demolition of old buildings and construction of new ones dominated in many countries. It was a policy that often led to displacement of the residents depending on what kind of housing was built. When social housing was provided in the city there was still an option for families with lower incomes to stay. From the beginning of the 1970s a change appeared in the renewal programmes which now put more weight on preservation of historic buildings and neighbourhoods (Skifter Andersen and Leather 1999).

In this context the central question is to what extent public urban renewal has resulted in a displacement of ethnic minorities from older neighbourhoods in the central cities. Very little research has been conducted to determine answers to this question, but they must be connected to the extent and character of urban renewal, the degree of government involvement and the degree to which objectives on the consequences for current residents in the neighbourhoods have been important. It is to be expected that renewal resulting in gentrification and displacement of low-income groups to a great extent also have resulted in ethnic minorities being dispelled.

Urban renewal policies have varied significantly across European countries. Comparative research (Turkington and Watson 2015; Skifter Andersen and Leather 1999) has found that public policies for housing rehabilitation and urban renewal in European countries are very different. Problems of urban decay have differed significantly from country to country, as do the conditions for solving them. In every country, urban problems have appeared in a specific geographical, economic and political context, and the possible instruments for solving the problems were often linked to the actual national structure of public administration and policies. Moreover, even if urban problems were alike in character, divergent political priorities and objectives for urban renewal have led to quite different policies in practice.

A central discussion in many countries has been to what extent existing residents in urban renewal areas should remain in or leave the neighbourhoods, where two main views exist (Bailey and Robertson 1997). One view emphasises that urban renewal should consider existing residents in the areas and in this way concentrate on redistribution of housing opportunities for low-income or 'deprived' groups. In this view gentrification in the areas is seen as a negative outcome. The alternative view is that public investments in urban renewal should encourage economic growth and stimulate in-migration of higher income groups who improve the tax base of the municipalities and contribute with further investments in housing rehabilitation. It was believed that the benefits of this growth and improvement would 'trickle down' to those with low incomes. Gentrification was thus seen as a desirable and positive outcome (see the discussion in Bailey and Robertson 1997).

In spite of different starting points in European countries concerning urbanisation and housing stock, there has been a connection between the applied urban renewal policies and the general housing and welfare policies (Skifter Andersen and Leather 1999). Countries such as Denmark and Sweden, where housing was seen partly as a public good, also have tended to consider housing rehabilitation as such, and great importance has been attached to security and the influence of sitting residents. In other countries such as the UK, Norway and France, where housing policies were regarded mainly as a means of solving housing problems for the poorer sections of the population, housing rehabilitation policies were much more oriented towards selected people and dwellings in cases where renewal on market terms was not expected to take place without public support. In cities in Southern Europe a combination of liberalisation of rental housing and private housing renewal have to a large extent expelled low-income groups from the central parts of the cities (Arbaci and Malheiros 2010). The relatively sparse area-based public renewal programmes did not embody any actual socially inclusive scope (Delladetsima 2003), and their final outcome has been physical renovation and the outward expulsion of vulnerable groups towards more peripheral areas.

Planning systems and segregation

As discussed in the beginning of this chapter, both spatial inequality and ethnic and social segregation are influenced by the way public authorities intervene in the development of cities by controlling the use of land. Of special importance is to what extent different kinds of housing are distributed in space. A dispersal of housing for low-income groups will only take place if local authorities through planning and steering of land use make it possible to build low-income housing in places where marked forces would result in more expensive housing. It has, however, been shown that strong planning and control of land use does not always result in reduced segregation. Sometimes strong planning has resulted in high separation of housing tenures and strong segregation (Arbaci 2007).

It emerges from the theoretical planning literature (Nadin and Stead 2008; Dasi 2007) that spatial planning systems are deeply embedded in their socio-economic, political and cultural context. Moreover, planning systems may have a certain degree of path-dependency, such as the persistence of institutions and cultures. Planning systems seem to be connected to welfare state systems. Nadin and Stead identify four different categories of planning systems:

- **Comprehensive Integrated**: Austria, Denmark, Finland, Germany, the Netherlands
- **Land-use Regulation**: Ireland, United Kingdom (and Belgium)
- **Regional Economic**: France, Portugal (and to some extent Germany)
- **Urbanism**: Greece, Italy, Spain (and some extent Portugal)

The *Comprehensive Integrated* system is an approach that is conducted through a very systematic and formal hierarchy of plans from the national to the local level, which co-ordinate public sector activity across different sectors, but focus more specifically on spatial co-ordination than on economic development. It requires responsive and sophisticated planning institutions and mechanisms and considerable political commitment. Public sector investments are necessary in bringing about the realisation of the planning framework and are also the norm. The *Land-use Regional* planning tradition has a much narrower scope with the purpose of regulating land-use change. The aim of the *Regional Economic* approach is to let regional economic development conform to some overall idea formulated by a central agency, using powers and funds at its disposal. Under this approach spatial planning has a very broad meaning relating to the pursuit of wide social and economic objectives. Where this approach is dominant, central government inevitably plays an important role. There is some public intervention in urban development, but it is to a higher degree coming from central or regional governments and displays the objectives formulated at those levels, most often objectives of economic growth and development. The *Urbanism* tradition is a spatial planning style characteristic of the Mediterranean countries, which takes place on the local level through building regulations. Public planning is weak and seldom interferes with the use of land. Nadin and Stead (2008) demonstrate a strong correspondence between welfare state models and models of planning. In Chapter 8 we will describe in more detail how these kinds of planning systems are found in countries belonging to different welfare systems.

The spatial structure of housing markets is, however, not only dependent on public planning and intervention in the use of land but also on the character of institutions and markets that shape and change the urban structure. One way to analyse these dynamics is to identify different systems of housing provision and land supply that create mechanisms of spatial differentiation. Three such factors have been identified in previous research, namely *forms of land supply*, *forms of housing production* and *forms of housing promotion* (see Arbaci 2007; Barlow and Duncan 1994).

The *forms of land supply* are connected with who owns the land used for urban development and to what extent it is owned by public actors. How much is land development affected by speculative market objectives and what is the role exercised by local democratic bodies? Unfortunately, studies of residential segregation have rarely examined the role of land-use regulations (Nelson et al. 2004).

The *forms of housing promotion* concern who are the final owners and providers of the housing produced. A dividing line is made between private and public/non-profit sectors.

Finally, the *forms of housing production* are about the kinds of developers and building firms operating in the housing sector. The kind of developers that are dominating depends not least on the form of land supply and ownership. For instance, if land is controlled by a local authority, it is more difficult for developers to make a profit on buying and using land for development, so instead they have to base profit-making activities on the building process alone. Moreover, it is important whether the market is dominated by large builders and developers or by small-scale builders. Large firms and developers tend to build larger and more uniform housing areas, especially if they are operating in cities where the local authorities do not own the land and do not regulate land use much.

It is therefore decisive how planning and housing systems are combined. High ethnic segregation can thus either be a result of a combined liberal housing and planning system or of a strong segmented housing system combined with a planning system that separates tenures in space. A lower rate of segregation can be obtained if housing policies result in a low degree of ethnic segmentation of tenures or if a strong planning system results in spatial dispersal of segmented tenures. Chapter 8 describes how these conditions vary across countries with different kinds of welfare systems.

Area-based initiatives and social mix policies to reduce segregation

In pace with the rate at which low-income groups moved out of slums in the central cities to housing in the suburbs from the beginning of the 1970s, new social problems evolved, especially in large social housing estates. These estates were erected in the post-war period to solve the enormous housing shortage. They were often built as large low quality and monotonous estates, which proved to be less attractive for people who had a choice on the housing market. Self-perpetuating processes of segregation and deprivation occurred in these places (Skifter Andersen 2003) with increasing social problems, deterioration and vacant dwellings. In most countries governments launched so-called area-based initiatives to combat the development on the most deprived estates (Wassenberg 2013).

A key strategy in these initiatives was to improve the conditions which made the neighbourhoods less attractive and more stigmatised compared to ordinary neighbourhoods (Skifter Andersen 2002). Among the measures were things like physical renovation and embellishment, direct social support for socially weak groups and education, job training and other attempts to get employment to residents with the purpose of reducing visible social problems and improving social status. An approach was also to change the housing structure of the estates by demolition and conversion to owner-occupied housing or private renting. The objective was to obtain a greater degree of tenure mixing in the neighbourhoods, which could result in social and ethnic mixing by attracting more natives and middle-class residents.

It can be said that the purpose of the initiatives was to stop and reverse the processes of segregation that had occurred around these neighbourhoods. As a consequence of the emerged problems in deprived housing estates, some countries also formulated so-called social mixing policies for development of new urban neighbourhoods based on mixing of housing tenures (Andersson et al. 2010; Blanc 2010; Livingston et al. 2013).

A special case in this connection is the city of Helsinki in Finland. In Helsinki, a tenure mixing strategy has been implemented since the 1990s by enforcing area-based tenure quotas for the production of new dwellings (Dhalman and Vilkama 2009). Mixing of tenure types has taken place mainly at the neighbourhood level, but also within blocks and housing estates. Urban renewal programmes have also been used to diversify a homogeneous housing stock in some older residential districts. In addition, mixing policies have been implemented through practices of social mixing in social housing allocation. Those with the most urgent need for housing were normally given first priority, but this principle of prioritisation has sometimes been breached in order to prevent residential segregation. The policy has resulted in significantly lower ethnic segregation, despite a strong segmentation of the housing market, as will be shown in the next section comparing ethnic segregation in the four Nordic capitals.

A considerable amount of research has been conducted on the concept of social mixing in neighbourhoods, but most of it has been concerned with the impact on living conditions in the neighbourhoods and on so-called neighbourhood effects (Bond et al. 2011; Bolt et al. 2010). Others (Livingston et al. 2013; Andersson et al. 2010) have made a general examination of the connection between tenure mix and social mix in neighbourhoods in Britain and Sweden, respectively, but found a rather weak connection. They conclude that the spatial distribution of tenures across neighbourhoods only has a weak impact on social and ethnic segregation. This is, however, in conflict with a Danish study from Copenhagen (Skifter Andersen et al. 2000) where it was found that 68 per cent of the social segregation and 46 per cent of ethnic segregation across 50 municipalities in the capital region could be explained by differences in housing markets across municipalities. In the next section we reference a study of ethnic segregation in the capitals of four Nordic countries that also shows a strong influence of the separation of housing tenures in space on ethnic segregation.

The effects of area-based initiatives on segregation have also been uncovered, but only to a limited extent. The general conclusion from the research conducted in the 1990s was that the initiatives had stopped a negative development in the neighbourhoods but not resulted in a significant change in the composition of residents (Skifter Andersen 2002). One of the reasons could be that a low social status, and an image as a minority neighbourhood, still dominated the pattern of local moves, as described in Chapter 6 on selective moves, in the years after the efforts.

Only in the cases where a substantial share of the dwellings was demolished have there been considerable changes. Evaluations of urban restructuring in the Netherlands show that physical interventions do have some effect on residential segregation within the restructured neighbourhoods themselves (Bolt et al. 2009). These studies show that the proportion of ethnic minorities increases at a slower pace, and that the number of low-income households decreases faster. Since urban restructuring areas are characterised by an over-representation of ethnic minorities and low-income households, this implies that physical interventions reduce the level of segregation, albeit the effect is quite modest. However, urban restructuring only leads to a decrease in segregation when a substantial number of social rented dwellings are replaced by owner-occupied dwellings.

Moreover, looking only at developments within the restructured neighbourhoods themselves does not yield a complete picture of the effects of urban restructuring. It is also necessary to examine the effects of the displacement of former residents from the neighbourhoods. The evaluation of the efforts in the Dutch estate of Bijlmermeer showed that the displaced residents to a large extent settled in nearby neighbourhoods that experienced an increased concentration of low-income residents and ethnic minorities (Wassenberg 2013). Tieskens and Musterd (2013) studied what happened to the relocatees as a result of demolition of four estates in Amsterdam. They found that immigrants and large families ended up in neighbourhoods with relatively more ethnic minorities and poverty. In another study, Bolt et al. (2009) concluded that displacement generally does not lead to desegregation, because displaced households did not move to less segregated neighbourhoods. A more general survey of evaluations of the experiences with demolition of neighbourhoods in the Netherlands and the US (Kleinhans and Varady 2011) concludes that 'existing research shows that many relocatees do recluster in low-income areas not much better than the public or social housing sites they moved from'. The cautious conclusion from this relatively sparse research could be that area-based initiatives have had some, but only a limited, impact on ethnic segregation.

The importance of housing markets and urban policies for ethnic segregation in four Nordic capitals

In this section we compare ethnic segregation in four Nordic capitals, Copenhagen, Helsinki, Oslo and Stockholm, and examine how the separation of ethnic minorities from natives is connected with ethnic segmentation of the housing market and the extent to which housing tenures have been separated spatially. We then try to explain the differences between the cities by differences in their housing policies and planning practices. The section is based on Skifter Andersen et al. (2016).

We also examine what could be called *the social housing hypothesis*. Earlier research (e.g. Murie and Musterd 1996; Arbaci 2007) focused on the importance of the size and character of the social housing sector in cities in the so-called Social Democratic welfare states of countries as a major explanation of differences in segregation. Arbaci (2007) argued that the relatively high level of ethnic segregation among these countries, in spite of more uniform and less socially segmented housing markets, could be ascribed to the existence of large spatial concentrations of social housing with many immigrants. It was assumed that a marginalised social sector with high concentrations of low-income groups also will result in high ethnic segregation, but this depends highly on the size of the sector and to what extent it is concentrated on large estates in certain parts of the cities.

Housing and planning policies in the Nordic countries

The Nordic countries are countries with a high degree of universalism, strong state involvement and extensive public responsibility in different social policy areas, and with strong local authorities. This also applies to housing and urban policies. However, as described in the last section of Chapter 6, housing policies and housing markets vary greatly across Nordic countries. It was also shown that there are substantial differences between ethnic minorities' position on the housing market across the four countries.

Sweden and Finland have smaller income differences between residents in renting and ownership (Skifter Andersen et al. 2013) and thus have more unitary housing systems, but the rental market in Finland is more segmented because social housing has been a restricted sector for low-income groups. Norway and Denmark have a stronger social separation between renting and owning (Skifter Andersen et al. 2013) and thus more dual systems. The rental sector in Denmark is more unitary with a large social housing sector, which is not socially restricted. However, rent control in private renting results in surplus demand and queues for this form of tenure, which is unfavourable to immigrants, because they seldom have good contacts with landlords (Skifter Andersen 2012). So in a way, the rental market is dual for immigrants. In Norway, the rental sector, and especially the social housing sector, is quite small, which is the reason why it is more segmented. Large cooperative sectors both in Sweden and Norway form a bridge between renting and owning, which to some extent makes the housing systems more unitary.

All four cities have had much public ownership of land and strong planning systems, but they differ somewhat concerning planning objectives. In Stockholm local planning regimes 20 to 30 years ago tended to be more pro-active in relation to developers and construction firms than recently (Andersson and Magnusson Turner 2014). In Helsinki, there has since the 1970s been a conscious strategy to mix tenures spatially (Dhalmann and Vilkama 2009; Andersson et al. 2010), which has not so much been the case in the other cities, especially not prior to 1980 when large parts of the suburbs were developed with large uniform housing areas. In Helsinki, Copenhagen and Stockholm, large social housing companies have developed large-scale housing. This was not found in Oslo, where social housing was developed by the municipalities in smaller-scale estates; but large cooperative companies exist in both Oslo and Stockholm that developed large housing estates. In Stockholm massive inner city conversions of social housing to cooperatives recently have led to a much stronger relative concentration of both private and public rental housing in the outer city, which makes inner city housing less available for low-income households, who have become more concentrated in the suburbs (Andersson and Magnusson Turner 2014).

The aim of the empirical analyses in this section is to examine the connection between ethnic segmentation of the housing market, the spatial structure of the housing market and ethnic residential segregation. The data used comes from four databases, one in each of the countries, containing register data on neighbourhoods of the four capitals, on their population and on their housing composition. Data is available on the ethnic composition of neighbourhoods and on the tenure composition of housing (for further description see Skifter Andersen et al. 2016).

Ethnic segregation in the four capital regions

As described in Andersson et al. (2010), Finland has received far fewer immigrants than the other countries, while the share of immigrants in Sweden is about 40 per cent higher than in Denmark and Norway. Moreover, the ethnic composition of immigrants varies between the countries and the capitals. In Finland the biggest Non-Western group comes from Eastern Europe, especially from Russia.

Finland has relatively few immigrants from countries outside Europe compared with the three other countries, where a total of 5 per cent of the whole population are immigrants coming from outside Europe. Stockholm in particular has many immigrants from non-European countries.

Table 7.1 shows segregation indices of dissimilarity for the capitals, which are measures for how much ethnic minorities are separated from natives. The figures are compared with the average for the four cities to illustrate how they differ from each other. Moreover, the level of immigration in the cities and an index of housing market segmentation, which measures the extent to which immigrants are unevenly distributed across housing tenures, are shown (see how the indices are calculated in Skifter Andersen et al. 2016).

The highest level of segregation between immigrants and natives is found in Stockholm, while the other three cities show values around or lower than the average of the four cities or below that. In all the cities Non-Western immigrants are more segregated, the most in Stockholm and the least in Helsinki. For different reasons, the cities with the highest proportion of immigrants (especially Stockholm) also have the highest level of segregation. The first reason is that the measure of segregation, the index of dissimilarity, is influenced by the proportion of immigrants in a city, as discussed in Skifter Andersen et al. (2016). The second is that the risk of white flight and avoidance might be higher in cities with a high representation of ethnic minorities.

The table displays considerable differences in ethnic segmentation across the capitals. Oslo has many immigrants living in owner-occupied housing and few in social housing, which is a very small sector. The distribution of immigrants in

Table 7.1 Ethnic spatial segregation of immigrants 20+ years (index of dissimilarity), the proportion of immigrants in the cities and indices of ethnic segmentation of housing markets

The capital regions	*Copenhagen*	*Helsinki*	*Oslo*	*Stockholm*	*Average for the cities*
Index of dissimilarity					
All immigrants	.25	.21	.27	.33	.27
Non-Western immigrants	.36	.27	.37	.48	.37
Proportion of immigrants %					
All immigrants	11.7	8.8	14.3	21.3	14.0
Non-Western immigrants	6.8	3.6	9.3	11.1	7.7
Index of ethnic tenure segmentation					
All immigrants	21	34	24	20	25
Non-Western immigrants	33	42	30	32	34

Source: Skifter Andersen et al. 2016

Note: The segregation indices for Copenhagen and Stockholm can be expected to be somewhat higher just because of smaller neighbourhood size.

Helsinki differs most from that of the natives and the index of tenure segmentation is high, especially for Non-Western immigrants. This is especially due to a high over-representation of immigrants in social housing. Also in Copenhagen and Stockholm, immigrants are over-represented in social housing, but at a lower level. The indices in these cities are still high, but lower than in Helsinki.

It is seen that there is not a straightforward connection between ethnic segmentation and segregation. The strongest segmentation is found in Helsinki, where the lowest segregation is found. Stockholm has the same segmentation as Copenhagen and Oslo but considerably higher segregation. These patterns are mostly a result of different spatial patterns of how housing tenures are located in the cities, as will be shown in what follows.

The spatial distribution of housing tenures across neighbourhoods

An analysis has been made of the extent to which tenures are spatially separated in the capitals in the countries. Table 7.2 is calculated on data from the databases described previously. Inside each tenure form, we calculate an index showing to what extent the distribution of residents in the tenure deviates from all tenures across neighbourhoods. Moreover, a combined index is calculated as the sum of the weighted indices per tenure.

The table shows marked differences between the cities. Copenhagen and Stockholm seem to have a much more spatially segregated housing market than Oslo and Helsinki. In both cities, social housing has the most uneven distribution across neighbourhoods. In Copenhagen, the high segregation of private renting and cooperatives is mostly due to the fact that these tenures are concentrated in the older housing stock located in central Copenhagen.

In Helsinki, owner-occupied single family housing is much more unevenly distributed than the other tenures. Cooperatives constitute a very small sector without any importance for the combined index, but the rented tenures and owner-occupied flats are quite evenly distributed across neighbourhoods. It is thus remarkable that the relatively large social housing sector is more dispersed

Table 7.2 Measures of the uneven spatial distribution of different housing tenures across neighbourhoods in four Nordic capitals, 2008

	Copenhagen	*Helsinki*	*Oslo*	*Stockholm*	*Average for cities*
Owner-occupied single family	.56	.62	.32	.55	.51
Owner-occupied flats	.47	.12	.39		.33
Cooperatives	.50	.57	.43	.36	.47
Private renting	.42	.27	.21	.45	.34
Social housing	.56	.33	.35	.52	.44
Combined index	.51	.23	.33	.46	.38

Source: Skifter Andersen et al. 2016

in Helsinki than in Stockholm and Copenhagen. This can be directly attributed to the fact that Helsinki has pursued a policy, launched already at the end of the 1970s, to mix different tenures in housing areas (Andersson et al. 2010; Dhalmann and Vilkama 2009). In Oslo, both owner-occupied housing and private renting are fairly evenly distributed across neighbourhoods. The tenure most spatially concentrated in Oslo is the cooperative sector, which comprises many large cooperative estates.

Estimating the connection between tenure segmentation and spatial segregation

Statistical models (linear regression) of the connection between tenure composition in neighbourhoods and the ethnic composition of them are used, in each city, to estimate an 'expected' number of immigrants in neighbourhoods based on the composition of tenures. If the expected number matches the real value across all neighbourhoods, ethnic segregation would be fully explained by the tenure composition.

The statistical analysis is based on data on each neighbourhood concerning the number of immigrants and the number of residents 20+ years old living in different tenures (dwellings). Four models were constructed, one for each city. The dependent variable is the proportion of immigrants in the neighbourhood. The models are described in Skifter Andersen et al. (2016). One of the independent variables is the average income in the neighbourhood to account for the effects of social segregation. We made this analysis in two steps. First, we built a regression model to estimate the connection between the tenure composition and the proportion of immigrants in neighbourhoods. We did this both for all immigrants and for the subcategory of Non-Western immigrants. In the second step, we used this model to estimate the expected tenure dependent number of immigrants in the neighbourhoods and then calculated the expected indices of dissimilarity for these expected numbers.

Figure 7.1 shows some results from the regression of the connection between tenure composition of neighbourhoods and the proportion of Non-Western immigrants that illuminates the importance of different tenures for ethnic segregation in the cities. The figure depicts the Standardised Beta value for each tenure in each city, produced in the regression.

It appears from the figure that social housing is most important for the presence of Non-Western immigrants in neighbourhoods in Helsinki, Copenhagen and Stockholm. In Oslo it is the location of cooperatives that is most important, followed by owner-occupied flats and social housing. The location of private renting also has some importance in Helsinki and Stockholm but not in Copenhagen.

The tenure models explain a considerable part of the variation in the proportion of immigrants in the neighbourhoods. The first row of Table 7.3 shows the R^2 values for the models for all immigrants. In all models more than 40 per cent of the variation is explained. The model for Helsinki has an even higher degree of explanation.

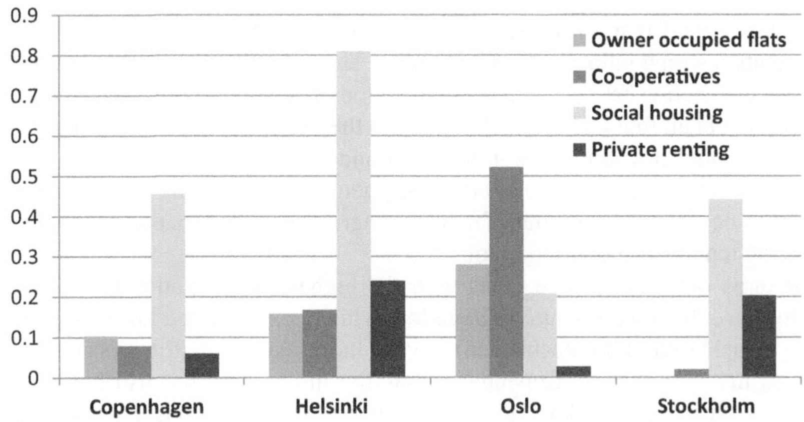

Figure 7.1 The influence of tenures on the proportion of Non-Western immigrants in neighbourhoods. Standardised Beta values from linear regression

Source: Skifter Andersen et al. 2016

Table 7.3 Estimated expected indices of segregation of immigrants computed from linear regression models, compared with the actual measured indices

	Copenhagen	*Helsinki*	*Oslo*	*Stockholm*
Variation explained by regression model: R^2 (all immigrants)	.44	.63	.43	.55
		Index of dissimilarity		
All immigrants				
Actual index	.25	.21	.26	.33
Estimated tenure dependent index	.15	.19	.17	.19
Tenure effect on segregation	58%	89%	65%	59%
Non-Western immigrants				
Actual index	.36	.27	.40	.48
Estimated tenure dependent index	.17	.23	.26	.29
Tenure effect on segregation	48%	83%	65%	61%

Source: Skifter Andersen et al. 2016

Note: The indices for Copenhagen and Stockholm can be expected to be somewhat too high due to smaller neighbourhood size.

The calculated expected indices can be interpreted as the segregation of immigrants that would have occurred if the housing market (tenure distribution) was the only cause of segregation. As expected, all the calculated expected indices are smaller than the actual indices. Yet the difference between the actual and the calculated indices varies across cities. The percentage of the segregation that in principle can be explained by the neighbourhoods' tenure composition is shown in the table labelled the 'Tenure effect on segregation'. It varies between 48 and 83 per cent for Non-Western immigrants.

Some conclusions from the Nordic study

We have found that an uneven spatial distribution of different housing tenures has quite a strong influence on segregation. The analyses have shown a relatively strong connection between tenure composition of neighbourhoods and ethnic concentration in all four cities, which illustrates the importance of the housing market for the spatial distribution of immigrants and for their separation from natives. However, the general degree of ethnic segmentation of housing markets does not seem to be the most important factor for segregation. The spatial distribution of housing tenures is more important.

It turns out that the two cities with the highest segmentation (Helsinki and Oslo) have the lowest tenure segregation, while those with the lowest segmentation (Stockholm and Copenhagen) have a more uneven distribution of tenures across neighbourhoods. Copenhagen has the most unevenly distributed social housing sector, followed by Stockholm, while the distribution is much more even in Helsinki and Oslo. Explanations for these patterns must be found in the systems of housing provision and land supply in the cities. In Copenhagen and Stockholm, large social housing estates were constructed in the period 1960 to 1975, which separated social housing from other tenures. The city of Helsinki has, from the 1970s onwards, pursued a conscious policy of mixing social housing into different neighbourhoods (Andersson et al. 2010; Dhalmann and Vilkama 2009). Moreover, private renting in Helsinki is also more dispersed than in Copenhagen and Stockholm. Oslo is a special case. Social housing is scattered in the city, but this small sector does not contribute much to the general measure of tenure segregation. The important sector for immigrants is the cooperatives, and this is the most segregated tenure in the city.

We expected that segregation would be influenced by a combination of high ethnic segmentation of the housing market and strong spatial segregation of tenures. The interesting thing is that the lowest segregation is found in the city (Helsinki) with the highest segmentation, but the lowest tenure segregation. The regressions of the connection between tenure compositions of neighbourhoods and the spatial distribution of immigrants also showed that the strongest connection is found in Helsinki. This indicates that ethnic segregation can be reduced by concentrating immigrants in a certain housing tenure if a spatial mixing policy of tenures is pursued. An explanation of the low segregation in Helsinki could partly be, however, that self-perpetuating processes of white flight and avoidance are weaker or at least concentrated in far fewer neighbourhoods.

Given the low ethnic segmentation of the housing market in Oslo, combined with the relatively low rate of segregation of tenures, one would have expected that segregation in Oslo would be much lower than in Copenhagen and Stockholm. However, Oslo has the same level of segregation as Copenhagen. Oslo also has higher concentrations of immigrants in some neighbourhoods than Copenhagen, and they are mostly found in neighbourhoods with cooperative housing. The most important tenure factor behind segregation thus is cooperatives built in larger estates in the suburbs, in particular in the northeast of Oslo.

Arbaci (2007) assumed that much ethnic segregation in the Nordic countries might be ascribed to the existence of large spatial concentrations of socially marginalised and segmented social housing, what could be called *the social housing hypothesis*. This study shows a much more differentiated picture. The hypothesis can only partly be confirmed for Copenhagen and Stockholm, where high concentrations of immigrants are found on social housing estates that are segregated from other tenures. In none of these cities, however, is social housing a marginalised sector restricted for the poor. It is accessible for all income groups and can thus also be chosen by other immigrants than the most deprived. This seems to have created a basis for stronger processes among immigrants of self-selected spatial concentration, which in turn might have reinforced white flight and avoidance. The case of Oslo shows that other tenures, cooperatives, can be the basis for ethnic segregation.

Note

1 This chapter is partly based on the article Skifter Andersen et al. (2016), written together with Roger Andersson, Institute for Housing and Urban Research at Uppsala University, Terje Wessel, Institute for Sociology and Geography at University of Oslo and Katja Vilkama at City of Helsinki Urban Facts.

References

Alba R. D. and Logan J. R. (1992) Assimilation and Stratification in the Homeownership Patterns of Racial and Ethnic Groups, *International Migration Review*, 26(4), pp. 1314–1341.

Alba R. D., Logan J. R., Stults B. J., Marzan G. and Zhang W. (1999) Immigrant Groups in the Suburbs: A Reexamination of Suburbanization and Spatial Assimilation, *American Sociological Review*, 64(3), pp. 446–460.

Allen J. (2004) Welfare Systems in Southern Europe. In Allen J. et al. *Housing and Welfare in Southern Europe*. Oxford: Blackwell.

Andersson R. and Magnusson Turner L. (2014) Segregation, Gentrification, and Residualisation: From Public Housing to Market Driven Housing Allocation in Inner City Stockholm, *International Journal of Housing Policy*, 14(1), pp. 3–29.

Andersson R. et al. (2010) *Immigration. Housing and Segregation in the Nordic Welfare States*. Department of Geosciences and Geography University of Helsinki.

Arbaci S. (2007) Ethnic Segregation, Housing systems and Welfare Regimes in Europe, *International Journal of Housing Policy*, 7(4), pp. 401–433.

Arbaci S. and Malheiros J. (2010) De-Segregation, Peripheralisation and the Social Exclusion of Immigrants: Southern European Cities in the 1990s, *Journal of Ethnic and Migration Studies*, 36(2), pp. 227–255.

Atkinson R. and Kintrea K. (2000) Owner Occupation, Social Mix and Neighbourhood Impacts, *Policy and Politics*, 28, pp. 93–108.

Bailey N. and Robertson D. (1997) Housing Renewal, Urban Policy and Gentrification, *Urban Studies*, 34, pp. 561–578.

Balchin P. (1996) *Housing Policy in Europe*. London: Routledge.

Barlow J. and Duncan S. (1994) *Success and Failure in Housing Provision. European Systems Compared*. Oxford. Elsevier Science.

Bolt G., Phillips D. and Van Kempen R. (2010) Housing Policy, (De)segregation and Social Mixing: An International Perspective, *Housing Studies*, 25(2), pp. 129–135.

Bolt G., Van Kempen R. and Van Weesep J. (2009) After Urban Restructuring: Relocations and Segregation in Dutch Cities, *Tijdschrift voor Economische en Sociale Geografie*, 100(4), pp. 502–518.

Bond L., Sautkina E. and Kearns A. (2011) Mixed Messages about Mixed Tenure: Do Reviews Tell the Real Story? *Housing Studies*, 26(1), pp. 69–94.

Blanc M. (2010) The Impact of Social Mix Policies in France, *Housing Studies*, 25(2), pp. 257–272.

Briggs X. (Ed.) (2005) *The Geography of Opportunity*. Washington, DC: Brookings Institution Press.

Castells M. (1989) *The Informational City: Information Technology, Economic Restructuring and the Urban – Regional Process*. Oxford: Blackwell.

Clark W. A. V. and Patel S. (2005) *Residential Choices of the Newly Arrived Foreign Born: Spatial Patterns and the Implications for Assimilation*. On-Line Working Paper Series, California Center for Population Research, UC Los Angeles, Los Angeles.

Dasi F. J. (2007) *Governance of Territorial and Urban Policies from EU to Local Level*. Final Report of ESPON Project 2.3.2. ESPON Coordination Unit, Esch-sur-Alzette.

Delladetsima P. (2003) The Olympic Village: A Redevelopment Marathon in Greater Athens. In Moulaert F., Rodrıguez A. and Swyngedouw E. *The Globalized City: Economic Restructuring and Social Polarization in European Cities*, pp. 65–89. Oxford: Oxford University Press.

Dhalmann H. and Vilkama K. (2009) Housing Policy and the Ethnic Mix in Helsinki, Finland: Perceptions of City Officials and Somali Immigrants, *Journal of Housing and the Built Environment*, 24, pp. 423–439.

Galster G. (2007) Neighbourhood Social Mix as a Goal of Housing Policy: A Theoretical Analysis, *European Journal of Housing Policy*, 7(1), pp. 19–43.

Hamnett C. and Butler T. (2010) The Changing Ethnic Structure of Housing Tenures in London, 1991–2001, *Urban Studies*, 47(1), pp. 55–74.

Kemeny J. (1995) *From Public Housing to Social Renting: Rental Policy Strategy in Comparative Perspective*. London: Routledge.

Kesteloot C. and Cortie C. (1998) Housing Turks and Moroccans in Brussels and Amsterdam: The Difference between Private and Public Markets, *Urban studies*, 35(10), pp. 1835–1853.

Kleinhans R. and Varady D. (2011) Moving Out and Going Down? A Review of Recent Evidence on Negative Spill over Effects of Housing Restructuring Programmes in the United States and the Netherlands, *International Journal of Housing Policy*, 11(2), pp. 155–174.

Knox P. (1995) *Urban Social Geography*. London: Longman.

Livingston M., Kearns A. and Bailey N. (2013) Delivering Mixed Communities: The Relationship between Housing Tenure Mix and Social Mix in England's Neighbourhoods, *Housing Studies*, 28(7), pp. 1056–1080.

Malheiros J. M. (2002) Ethni-Cities: Residential Patterns in Northern European and Mediterranean Metropolis. Implication in Policy Design, *International Journal of Population Geography*, 8(2), pp. 107–134.

Murie A. and Musterd S. (1996) Social Segregation, Housing Tenure and Social Change in Dutch Cities in the Late 1980s, *Urban Studies*, 33(3), pp. 495–516.

Musterd S. and Andersson R. (2005) Social Mix and Social Opportunities, *Urban Affairs Review*, 40(6), pp. 761–790.

Musterd S. and Ostendorf W. (1998) Segregation, Polarisation and Social Exclusion in Metropolitan Areas. In Musterd S. and Ostendorf W. *Urban Segregation and the Welfare State. Inequality and Exclusion in Western Cities*, pp. 1–14. London: Routledge.

Nadin V. and Stead D. (2008) European Spatial Planning Systems, Social Models and Learning, *The Planning Review*, 44, pp. 35–47.

Nelson K. P. (1988) *Gentrification and Distressed Cities*. Madison, WI: University of Wisconsin Press.

Nelson A. C., Dawkins C. J. and Sanchez T. W. (2004) Urban Containment and Residential Segregation: A Preliminary Investigation, *Urban Studies*, 41(2), pp. 423–439.

Park R. E. (1925) The City: Suggestions for the Investigation of Human Behaviour in the Urban Environment. In Park R. E. and Burgess E. W. *The City*, pp. 1–46. Chicago, IL: University of Chicago Press.

Park R. E. and Burgess E. (1969) *Introduction to the Science of Sociology*, 1921 Reprint. Chicago, IL: University of Chicago Press.

Sassen S. (1994) *Cities in a World Economy*. London: Pine Forge Press.

Skifter Andersen H. (2012) *Housing Policies in the Nordic Countries and Its Implications for the Housing of Immigrants*. Horsholm: Danish Building Research Institute.

Skifter Andersen H. (2010) Spatial Assimilation in Denmark. Why Do Immigrants Move to and from Multi-ethnic Neighbourhoods? *Housing Studies*, 25(3), pp. 281–300.

Skifter Andersen H. (2003) *Urban Sores. On the Interaction between Segregation, Urban Decay and Deprived Neighbourhoods*. London: Ashgate Publishers.

Skifter Andersen H. (2002) Can Deprived Housing Areas be Revitalised? Efforts against Segregation and Neighbourhood Decay in Denmark and Europe, *Urban Studies*, 39(4), pp. 767–790.

Skifter Andersen H. (1998) Gentrification or Social Renewal? Effects of Public Supported Housing Renewal in Denmark, *Scandinavian Housing and Planning Research*, 15(3), pp. 111–128.

Skifter Andersen H., Andersen H. T. and Ærø T. (2000) Social Polarisation in a Segmented Housing Market. Segregation in Greater Copenhagen, *Geografisk Tidsskrift*, 100(1), pp. 71–83.

Skifter Andersen H., Andersson R., Wessel T. and Vilkama K. (2016) The Impact of Housing Policies and Housing Markets on Ethnic Spatial Segregation: Comparing the Capital Cities of Four Nordic Welfare States, *International Journal of Housing Policy*, 16(1), pp. 1–30.

Skifter Andersen H. and Leather P. (Ed) (1999) *Housing Renewal in Europe*. Bristol: The Policy Press.

Skifter Andersen H., Magnusson Turner L. and Søholt S. (2013) The Special Importance of Housing Policy for the Housing Situation of Immigrants. Evidence from a Comparison of Four Nordic Countries, *International Journal of Housing Policy*, 13(1), pp. 17–36.

Soja E. (1980) The Socio-spatial Dialectic, *Annals, Association of American Geographers*, 70, pp. 207–225.

South S. J. and Crowder K. D. (1997) Escaping Distressed Neighborhoods: Individual, Community, and Metropolitan Influences, *American Journal of Sociology*, 102, pp. 1040–1084.

South S. J., Crowder K. and Pais J. (2011) Metropolitan Structure and Neighborhood Attainment: Exploring Intermetropolitan Variation in Racial Residential Segregation, *Demography*, 48(4), pp. 1263–1292.

Tieskens K. F. and Musterd S. (2013) Displacement and Urban Restructuring in Amsterdam; Following Relocatees after Demolition of Social Housing, *Urban Research & Practice*, 6(2), pp. 194–210.

Turkington R. and Watson C. (Eds) (2015) *Renewing Europe's Housing*. Bristol: Policy Press.

Van Ham M., Manley D., Bailey N., Simpson L. and Maclennan D. (2013) Understanding Neighbourhood Dynamics: New Insights for Neighbourhood Effects Research. In Van Ham M., Manley D., Bailey N., Simpson L. and Maclennan D. *Understanding Neighbourhood Dynamics: New Insights for Neighbourhood Effects Research*. New York: Springer.

Van Kempen E. T. (1994) The Dual City and the Poor: Social Polarisation, Social Segregation and Life Chances, *Urban Studies*, 31(7), pp. 995–1015.

Van Kempen R. (2003) *Segregation and Housing Conditions of Immigrants in Western European Cities*. Eurex Lecture 7. 13 March. Website: at www.shiva.uniurb.it/eurex/syllabus/lecture7/Lecture7-VanKempen.pdf.

Van Kempen R. and Murie A. (2009) The New Divided City: Changing Patterns in European Cities, *Tijdschrift voor Economische en Sociale Geografie*, 100(4), pp. 377–398.

Wassenberg F. (2013) *Large Housing Estates: Ideas, Rise, Fall and Recovery*. Delft: Delft University Press.

Wessel T. (2000) Social Polarisation and Socioeconomic Segregation in a Welfare State: The Case of Oslo, *Urban Studies*, 37(11), pp. 1947–1967.

Wright R., Ellis M. and Parks V. (2005) Re-placing Whiteness in Spatial Assimilation Research, *City and Community*, 5(2), pp. 111–135.

Zwiers M., Kleinhans R. and van Ham M. (2015) *Divided Cities: Increasing Socio-Spatial Polarization within Large Cities in the Netherlands*. IZA DP No. 8882. Forschungsinstitut zur Zukunft der Arbeit, Bonn.

8 Ethnic segregation and its causes across European welfare states

In the previous chapters we have discussed the causes of ethnic segregation and how it is influenced by different conditions in countries. In this chapter we will, based on available data, look at how these conditions vary across European countries; especially how they are connected to typified political systems, so-called welfare state systems. By comparing countries we will try to unravel how these conditions influence each other and the extent of segregation in countries.

Welfare state types

Ethnic spatial segregation differs greatly from country to country, to some extent also from city to city within the same country. It is connected to social segregation, but not in a consistent way. Social inequality is a driving force in segregation and the causes of social segregation are also important explanations for the spatial separation of ethnic minorities in cities, as it is a common fact that ethnic minorities, especially newly arrived immigrants, have lower incomes than the native majority in a country, and that such differences have an effect on segregation of minorities.

Social inequality and social segregation depend on the character of labour markets and welfare systems, which thus also might be important determinants of ethnic segregation. In this context integration policies and the extent to which ethnic minorities have been able to become economically integrated as well as obtain education, jobs and good incomes, have influenced their ability to choose better neighbourhoods and mix with the native majority.

Welfare regimes can, in multiple ways, affect and shape the patterns of segregation. Income redistribution, citizenship, access to education, health care, etc. are all welfare factors that have an influence on immigrants' chances of having success in their new country and on their ability to integrate and to choose by themselves where they want to settle. Housing systems, especially the extent of social housing, affect the housing options of ethnic minorities, and the location of different kinds of housing often determines where they can settle.

West European countries have in the literature been divided into different kinds of welfare state systems based on the original contribution by Esping Andersen (1990). Mostly they have been divided into four groups (Arbaci 2007):

1. Social Democratic welfare cluster (Sweden, Denmark, Finland, Norway, Netherlands),
2. Corporatist welfare cluster (Germany, Austria, France, Belgium),
3. Liberal welfare cluster (UK, Ireland, Switzerland, Luxembourg and to some extent also Belgium),
4. Latin Rim welfare cluster (Greece, Portugal, Spain, Italy, Cyprus, Malta).

In this section we will moreover make a group of the new EU members from Eastern Europe.

Social Democratic welfare states have universalistic systems that promote an equality of high standards, rather than an equality of minimal needs. This implies publicly supplied welfare services, to reduce the social division introduced by market-based access to welfare services, as well as pre-emptively socialising the costs of caring for children, the aged and the helpless, rather than waiting until the family's capacity to support them is depleted.

Liberal welfare states are characterised by modest, means-tested assistance, and targeted at low-income recipients. Their strict entitlement rules are often associated with stigma. This type of welfare state encourages market solutions to social problems – either passively, by guaranteeing only a minimum, or actively, by directly subsidising private welfare schemes.

Corporatist and Latin Rim welfare systems are typically shaped by traditional family values, and tend to encourage family-based assistance dynamics. Social insurance in this model typically excludes non-working wives, and family benefits encourage motherhood. State assistance will typically only step in when the family's capacity to aid its members is exhausted. In the Latin Rim systems, income polarisation is counterbalanced by Christian social policy (family, patrimony) and informal access to employment and resources.

There is a geographical dimension in the division of countries into welfare systems, as the Social Democratic welfare cluster primarily consists of countries from Northern Europe, the Corporatist and Liberal clusters of countries from Central and most Western Europe, and the Latin Rim cluster is located to the south. There are also some significant differences in the economic prosperity across clusters, which might influence ethnic segregation. Figure 8.1 shows the average GDP per inhabitant in the EU countries and uncovers considerable differences across countries and clusters.

Except for some smaller 'tax havens states' in the Liberal cluster, GDP is highest in the Social Democratic cluster closely followed by the Corporatist cluster and the UK. GDP is somewhat lower in the Latin Rim countries and much lower in Eastern Europe.

As ethnic minorities mostly have lower incomes than average, their situation is influenced not only by the general economic conditions, but also by the degree of social inequality. Income inequality varies quite a lot across the countries and welfare clusters, as can be seen from Figure 8.2.

In Western Europe, the general trend is that inequality is highest in the Latin Rim countries followed by the Liberal cluster and the Corporatist cluster. Equality

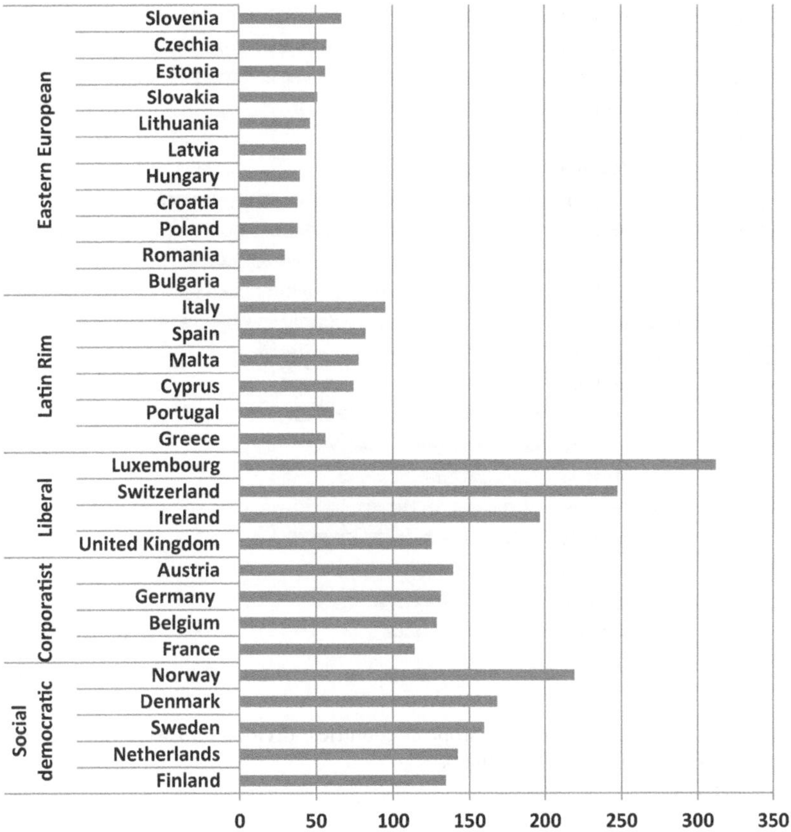

Figure 8.1 Average GDP per inhabitant in EU countries divided into welfare clusters, 2016

Source: EUROSTAT

is, as could be expected, highest in the Social Democratic cluster. Among the Eastern countries the situation is more diverse.

A general conclusion from these figures is that the economic conditions for low-income groups, and then also for ethnic minorities, are best in the Social Democratic countries and worst in the Latin Rim and Eastern Europe.

Ethnic minorities in European countries

Ethnic segregation might be influenced by the size and composition of ethnic minorities. The definition of ethnic minorities differs somewhat from country to country in Europe. In some countries only non-citizens are included. Others include all residents born outside the country and some also include the descendants of immigrants. In the EUROSTAT statistics only Non-European citizens are included.

It is difficult to find statistics that illuminate the extent of ethnic minorities across European countries. We know from some case studies (e.g. Musterd 2005;

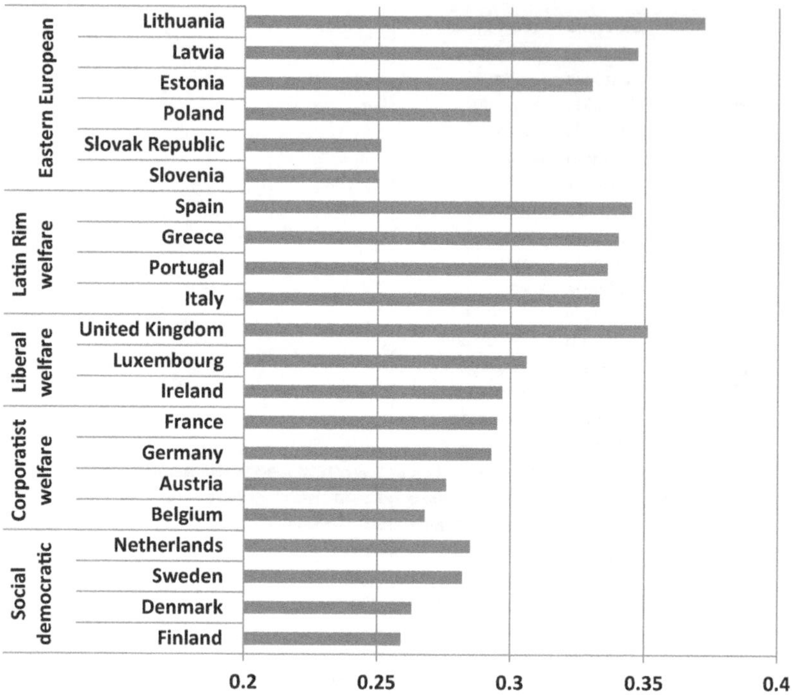

Figure 8.2 Income inequality in European countries, divided into welfare state clusters, measured by the Gini Index

Source: OECD 2017

Musterd and Van Kempen 2009) that the extent and composition of ethnic minorities diverge greatly from country to country. In former colonial power countries, immigrants from their colonies are often dominating, in others the character of labour import and refugees has had essential importance.

Figure 8.3 shows the proportion of residents in European countries who are not born in the country.

The proportion of foreign born varies greatly across countries, also within welfare clusters. The highest proportions of foreign born are found in the Liberal and the Corporatist clusters. The Social Democratic clusters have had lower and later immigration, except for Sweden. In the eastern part of EU, most countries have fewer ethnic minorities and they probably consist mainly of immigrants from Eastern Europe: Russia, Ukraine, etc.

Incomes of ethnic minorities

The ability of ethnic minorities to choose the housing they want and to decide where to live in the city depends on their economic resources and to what extent

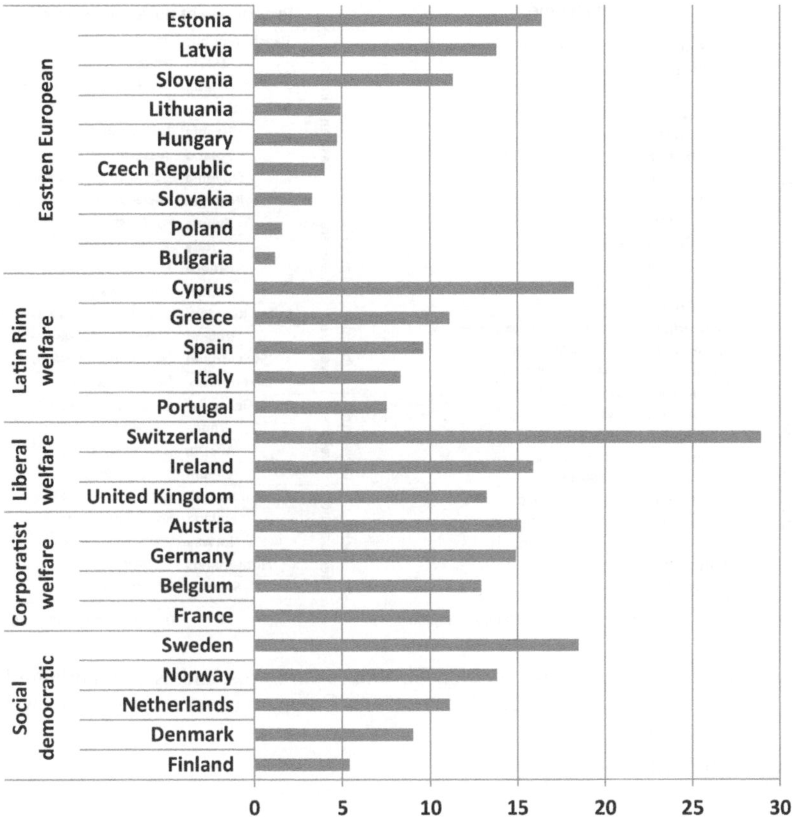

Figure 8.3 Proportion (%) of population who were not born in the country of residence, in European countries grouped by welfare state types

Source: United Nations Department of Economic and Social Affairs 2015 report: Trends in International Migrant Stock

their income is comparable with the native majority. The income of ethnic minorities varies from country to country in pace with the general income level, but not quite in a consistent way. Statistics on the median income for Non-EU citizens are shown in Figure 8.4 together with the proportion it makes up of the national medium income.

The figures might underestimate the incomes of ethnic minorities in general because it should be expected that Non-EU citizens have a lower income than ethnic minorities in general. Those who have obtained citizenship probably have higher incomes than those who have not. Citizens from more wealthy countries like the US and Norway are, however, also included. Some countries like the UK and Switzerland are known for having many wealthy residents from overseas countries.

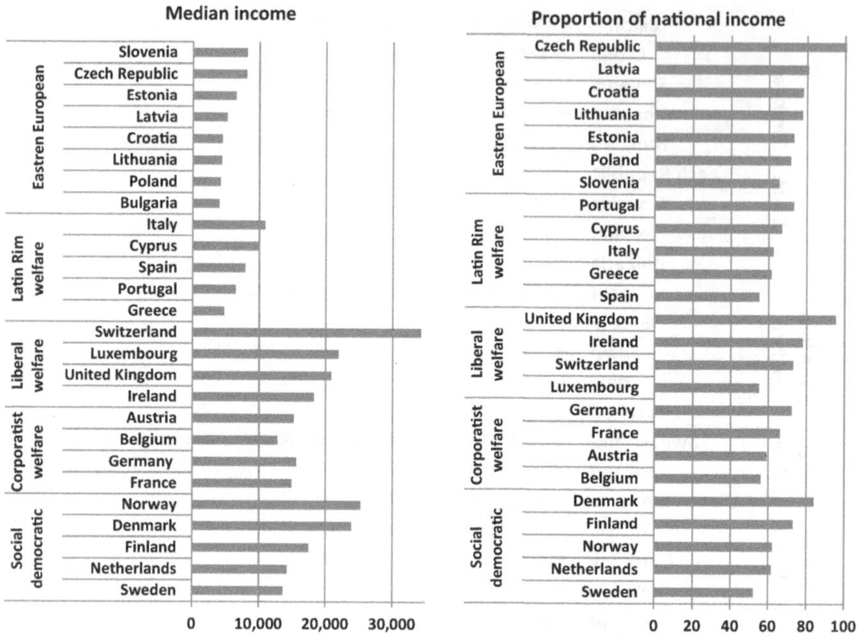

Figure 8.4 The median income for Non-EU citizens in Euros, and as a proportion of the national average median income (%) in European countries grouped by welfare state types

Source: EUROSTAT

It is seen from the left part of the figure that the incomes of Non-EU citizens vary greatly across countries, and from the right figure that this is not a simple consequence of different national income levels. In Sweden the median income of Non-EU citizens is only a little more than half of the national median income. In the UK they are nearly the same. The figures expose differences across countries on the extent to which ethnic minorities are less affluent than the average population. It is not possible to uncover major differences across welfare state clusters. Ethnic minorities in the Latin Rim countries seem to have a little lower income level compared to the national average. The low figure for Sweden could partly be explained by a higher rate of citizenship, whereas the remaining Non-EU citizens have somewhat lower incomes.

Ethnic segregation in Europe

While there are several studies on segregation in selected cities and countries, literature comparing many countries and ethnic groups is scarce, sometimes only comparing a few countries. The studies mostly concern the spatial segregation and concentration of specific ethnic groups. In many countries, however, ethnic minorities are concentrated in neighbourhoods containing several different groups, so-called multi-ethnic neighbourhoods. In these countries the most important kind of segregation is the separation of the whole group of ethnic minorities from the native majority.

A general conclusion is, however, that ethnic and social segregation in Europe, with a few exceptions, is lower than the segregation of the black population in North America, but at the same level as other ethnic minorities in these countries (Iceland et al. 2011; Musterd 2005). From the few comparisons of a larger number of cities it can also be concluded that ethnic segregation in Europe differs greatly across countries, cities and ethnic groups.

Musterd (2005) compared data from several studies on the segregation of selected ethnic groups in a number of cities and found that the most segregated groups were immigrants from south East Asia in British cities. Among the 20 most segregated groups, only three were from countries other than the UK; two from Belgium (Bosnians in Antwerp and Moroccans in Brussel) and one from Sweden (Iranians in Stockholm). These results are confirmed by Johnston et al. (2002), who examined ethnic segregation in all English cities and found particularly high concentrations of immigrants from countries like India, Bangladesh and Pakistan in a number of cities. In some neighbourhoods one ethnic group comprised more than 60 per cent of the residents. In the comparison of cities made by Musterd (2005) only a few cities from France and Southern Europe were included, but they all had relatively low segregation. This was also concluded by Malheiros (2002) in a comparison of segregation in four Southern-European cities with some Northern-European cities in Britain, the Netherlands and Belgium. The chosen cities from the north were, however, some of the most segregated cities, according to Musterd. Especially in Barcelona and Lisbon, quite high segregation of groups like Moroccans and immigrants from Cape Verde were found. The same was found by Musterd and Van Kempen (2009), who used some other data finding higher segregation in some Southern cities like Barcelona and Lisbon, but low segregation in others.

All the studies demonstrated large differences between different groups within the same city. Moreover, the same ethnic groups were not segregated to the same extent in different cities. Large differences existed for them across different countries but also between different cities within the same country. These results can be influenced by limitations on which cities have been examined and by differences in the segregation data used, but they indicate that ethnic background, national conditions and individual characteristics of cities all have an influence, but not a decisive one, on ethnic segregation.

Arbaci (2007) examined the connection between ethnic segregation of single groups in cities and factors like the relative size of the ethnic group in the city, the location of concentration areas (central city, suburbs, etc.), welfare state systems and housing markets in 30 European cities. She did not find that the extent of immigration had an importance for segregation of single groups. There was not a direct correlation between the proportion of an ethnic group (per cent of total population) and the level of spatial segregation. Moreover, the level of spatial segregation was unrelated to the population size of the cities. She found, however, that levels of spatial segregation of single groups differed across welfare clusters. Cities characterised by a liberal welfare regime scored the highest levels of segregation, especially in the city centre areas. At the other extreme, cities characterised by a Corporatist welfare regime, followed by cities with a Latin Rim regime, scored the lowest levels of segregation, and the areas of ethnic over-representation were located either in the inner city or in the inner suburbs. Cities within the Social Democratic cluster were an intermediate case, with the highest concentrations outside the city centre.

Glikman and Semyanov (2012) used data obtained from the European Social Survey (ESS) conducted in 2002 in 13 European countries, where ethnic minorities, defined as respondents who either were born outside the country or who have at least one parent that was born outside the country, were asked about their view on the ethnic composition of their neighbourhood. Figure 8.5 shows the proportion of the respondents who viewed their neighbourhood as dominated by ethnic minorities. This can be interpreted as the extent of the general spatial concentration of ethnic minorities in countries. The answers might be influenced by different perceptions of what is meant by 'dominated by ethnic minorities', and what is meant by the term neighbourhood.

It is seen that the answers vary greatly across countries, but there also seem to be some differences across welfare state clusters. The results are not quite

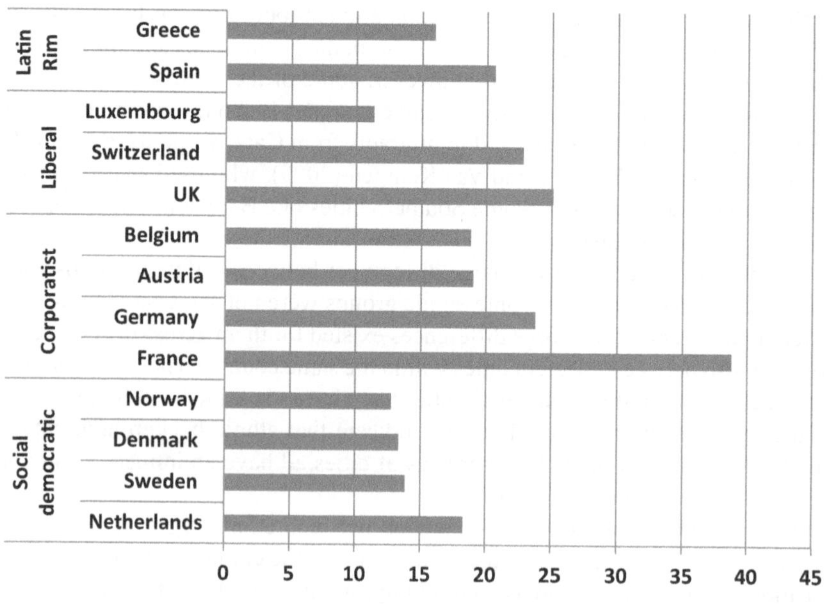

Figure 8.5 Percentage of ethnic minorities who answered that they lived in a neighbourhood dominated by ethnic minorities, 2002, in some Western European countries grouped by welfare state types

Source: Glikman and Semyanov 2012

in accordance with the evidence found by Musterd (2005) and Arbaci (2007). Corporatist welfare states seem to have a relatively high concentration of ethnic minorities. Most ethnic minorities in France view their neighbourhood as ethnic, followed by the UK and Germany. Scandinavian countries and to some extent Holland are at a lower level of ethnic concentration. Southern countries like Spain and Greece are measured to have medium segregation.

An important reason for the deviation between the data provided by Glikman and Semyonov and by Musterd and others is that the last studies concern the segregation of single ethnic groups, while the data of Glikman and Semyonov include the presence of all kinds of ethnic minorities in the neighbourhood. In some countries like Britain ethnic minorities tend to concentrate in neighbourhoods dominated by their own ethnic group, while in other countries they have settled in neighbourhoods with many different ethnic groups and with few residents belonging to the native majority. In this way we can talk about two kinds of ethnic segregation; segregation of single ethnic groups (from the native majority) and segregation between the ethnic majority and ethnic minorities in general. Figure 8.6 uses some data from the European social Survey 2014, where residents in 22 countries were asked about the ethnic composition of their neighbourhood. The possible answers were ‘Almost no minorities’, ‘Some minority groups’ and

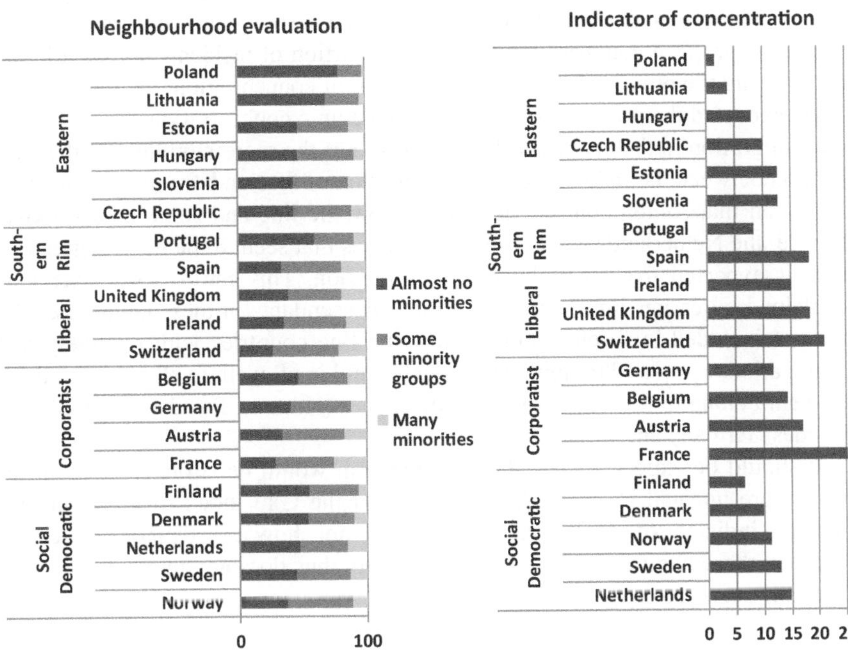

Figure 8.6 The extent of living in multi-ethnic neighbourhoods and a calculated indicator of ethnic concentration

Source: ESS European Social Survey 2014

'Many minorities'. The left side of the figure shows how the responses for different countries, grouped by welfare cluster, are distributed across these answers.

The left part of the figure displays to what extent the population experience living in a homogeneous neighbourhood or one that in varying degrees is composed of ethnic minorities. As demonstrated in Figure 8.5, for ethnic minorities Figure 8.6 shows large deviations across countries concerning the proportion of the whole population living in homogenous neighbourhoods, but not quite so significant differences across welfare state clusters except for the Eastern countries. Also in the Social Democratic cluster is it a little more common to live in a homogenous neighbourhood. The proportion living in neighbourhoods with many minorities varies greatly both inside and across welfare clusters.

The right part of the figure depicts an indicator of spatial concentration of ethnic minorities. It is calculated as the proportion of the residents not living in native homogenous neighbourhoods that are staying in neighbourhoods with many minorities. It shows to what extent neighbourhoods containing ethnic minorities are evaluated as multi-ethnic neighbourhoods. The figure shows a quite similar pattern as Figure 8.5, even if some countries inside the welfare clusters have changed position. Germany has thus moved to the position as the least segregated country in the Corporatist cluster.

We can use the data from the ESS (as in Figure 8.6) to examine to what extent segregation, measured by the calculated indicator, is connected to the proportion of ethnic minorities in the country. Figure 8.7 shows the connection across Western European countries between the proportion of residents not born in the country and the previously calculated indicator of concentration.

As expected there is a connection between ethnic concentration and the extent of immigration to the countries. The figure shows the regression line for the connection between the indicator and the proportion of foreign born in the countries. It is seen that some countries lie much below the line and some others above. France, the Netherlands and Spain are countries that seem to be more segregated than is expected from their level of immigration. This can to some extent be explained by earlier immigration and more descendants in these countries, who are not included in foreign born. Below the line countries with relatively less segregation, compared to immigration, are found in all welfare clusters, but three of them belong to the Social Democratic cluster (Norway, Sweden and Finland), besides Germany and Ireland.

It should be expected that ethnic segregation would be stronger in countries with more unequal incomes. A comparison of the Gini index for the countries with the indicator of ethnic concentration does not show a systematic connection between these two measures. Income inequality thus does not automatically lead to ethnic segregation.

Preferences for ethnic homogeneity or ethnic mix in neighbourhoods

As discussed in Chapter 5, preferences among the ethnic majority, and the extent to which they avoid moving to multi-ethnic neighbourhoods, are some of the

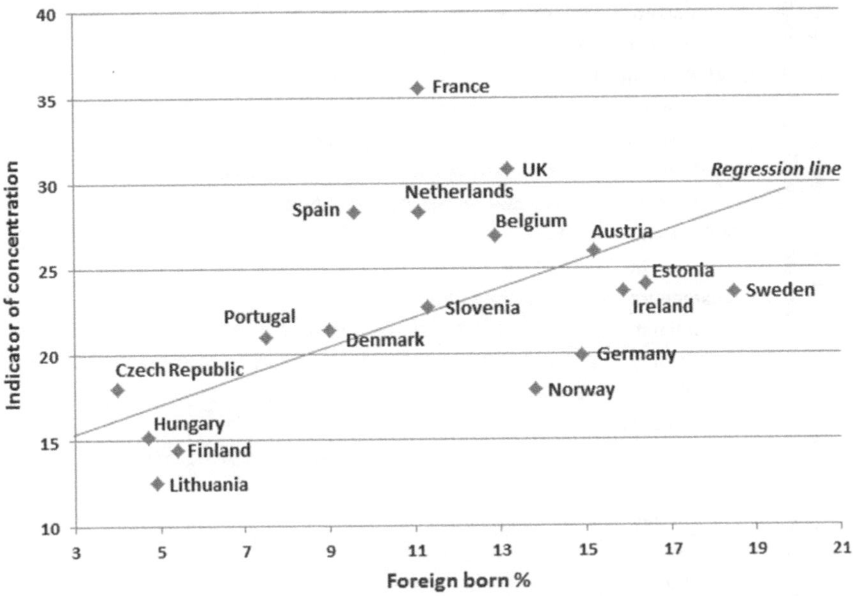

Figure 8.7 The connection across European countries between the proportion of residents not born in the country and an indicator of ethnic concentration (as in Figure 8.6)

Sources: United Nations Department of Economic and Social Affairs 2015 report: Trends in International Migrant Stock and ESS European social Survey 2014

major factors explaining ethnic segregation. In the 2002 European Social survey (ESS) from 20 European countries, residents were asked about their preferences for living in ethnic homogeneous or mixed neighbourhoods (the question was formulated as 'Suppose you were choosing where to live, which of the three types of areas on this card would you ideally wish to live in? An area where almost nobody was of a different race or ethnic group from most (country) people; some people were of a different race or ethnic group from most (country) people; many people were of a different race or ethnic group; or it would make no difference'). Figure 8.8 shows the proportion of respondents stating that they want to live in a homogeneous neighbourhood with 'Almost nobody of a different race or ethnic group'.

It is seen from the figure that preferences for living in ethnic homogeneous neighbourhoods vary greatly across countries and also inside welfare clusters. Preferences fluctuate from 45 per cent in Greece to 18 per cent in Switzerland. In the Social Democratic cluster, Sweden and Norway have the weakest preferences, in the Corporatist cluster it is Germany and France, and in the Liberal cluster the UK and Switzerland. All the countries in the Latin Rim and Eastern clusters, except for Italy, have higher preferences. Semyonov et al. (2007) conducted a

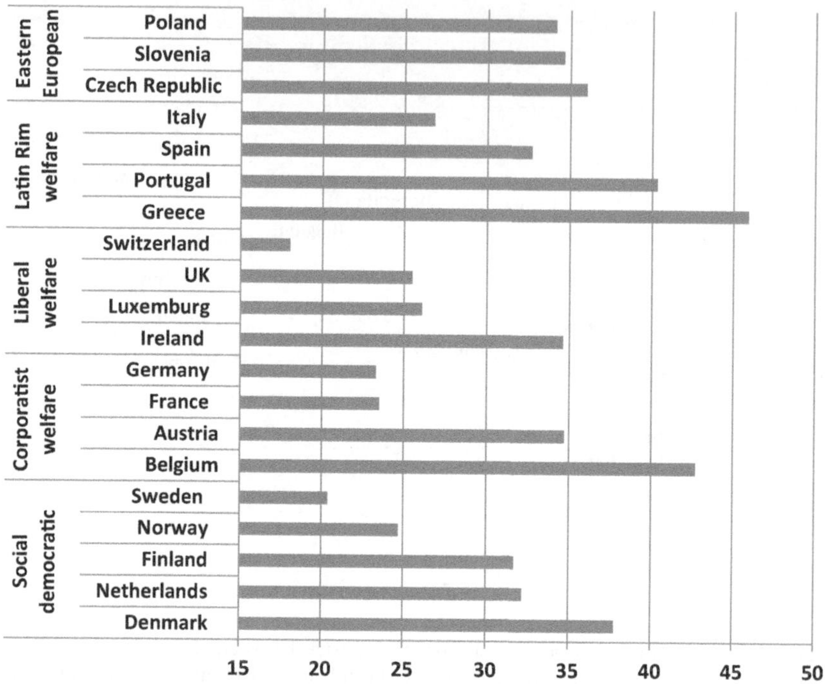

Figure 8.8 Proportion (%) of respondents answering that they want to live in a neighbourhood with 'Almost nobody minority race/ethnic group'

Source: European Social survey (ESS) 2002

statistical analysis on this data to control for differences in the social and demographic composition of the population, but found that this did not influence the differences across countries much.

The connection between attitudes towards ethnic minorities and preferences for homogeneity

As described in Chapter 5, the inclination among the native majority to avoid neighbourhoods with ethnic minorities can be a result of prejudices towards them. Good data on differences in attitudes towards ethnic minorities across European countries has been provided through the survey data from the 2002 European Social survey (ESS) from 20 European countries. Semyonov and Glikman (2009) used this data to illuminate differences across countries concerning anti-minority attitudes in the populations. They calculated what they called a *perception of threat index*, which was constructed from a couple of survey questions on fear of negative consequences of immigration like competition for jobs, public expenses, cultural concerns and crime. The results are shown in Figure 8.9.

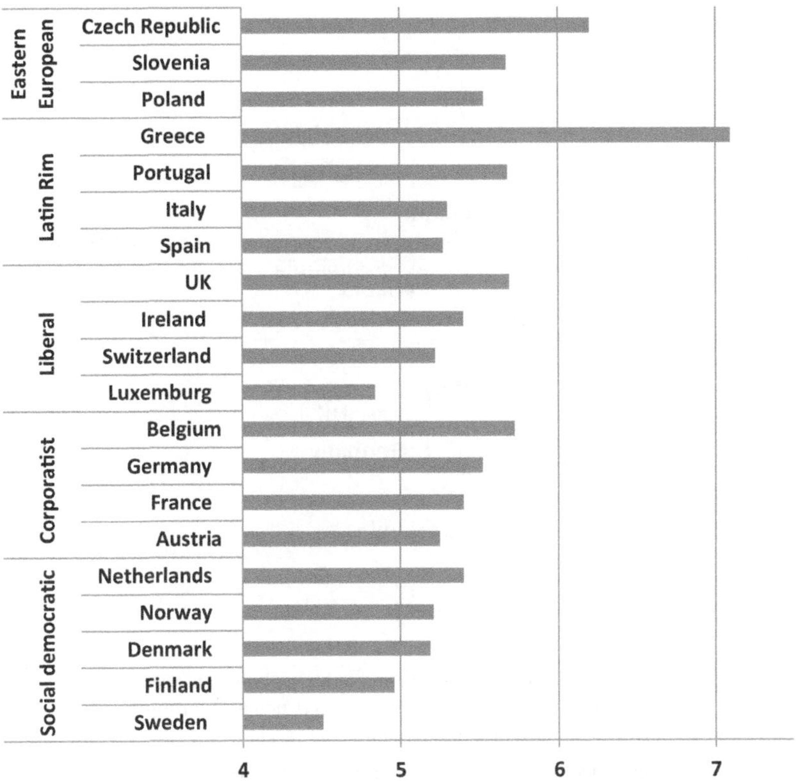

Figure 8.9 Negative attitudes towards ethnic minorities across European countries measured by a perception of threat index

Source: calculated by Semyonov and Glikman (2009) based on the 2002 European Social survey (ESS)

The figure shows pronounced differences across the countries on the extent of negative attitudes towards ethnic minorities. In contrast to the findings concerning preferences for homogeneity shown in Figure 8.8, the results in Figure 8.9 to some extent differ between welfare state clusters. The perception of threat index is the lowest in the Nordic Social Democratic countries. It is also low in Luxembourg and Switzerland, which can be explained by that these countries mostly have received immigrants from Western countries. In the upper scale is found a mix of countries belonging to the Liberal welfare cluster (UK and Belgium), the Latin Rim welfare cluster (Greece, Portugal) and some East European countries. Countries from the Corporatist welfare cluster (Germany Austria, France) are in general found in the middle of the order.

The data from the 2002 European Social survey makes it possible to illuminate the connection between preferences for ethnic homogeneity and negative attitudes towards ethnic minorities, which is shown in Figure 8.10.

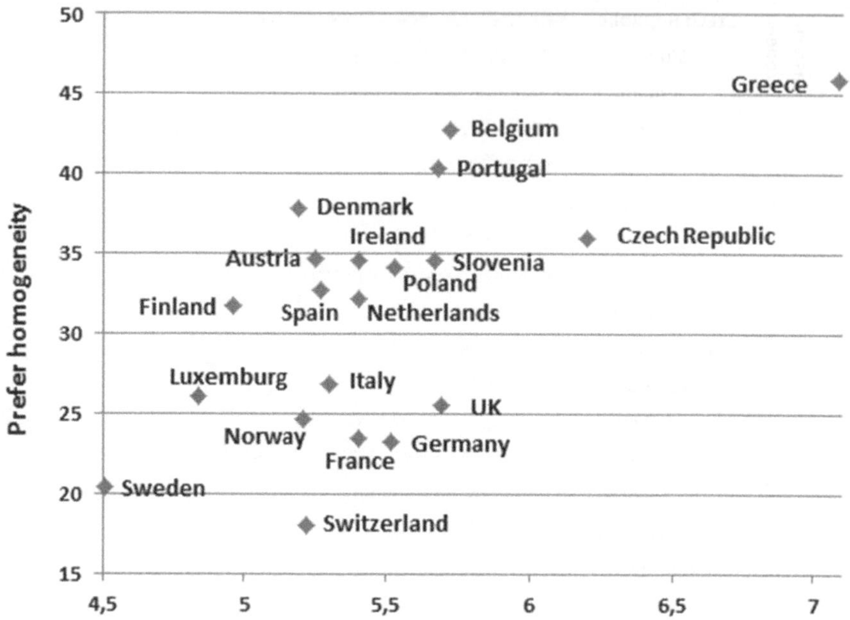

Figure 8.10 The connection across European countries between preferences for ethnic homogeneity (%) in neighbourhoods and negative attitudes towards ethnic minorities measured by the perception of threat index

Sources: Semyonov and Glikman 2009 and European Social survey (ESS) 2002

As can be seen from the figure, countries with a high perception of threat also tend to have stronger preferences for ethnic homogeneity in neighbourhoods. To the extent that these preferences lead to white avoidance, it can thus be concluded that ethnic segregation is influenced by specific political and economic conditions in countries that lead to negative attitudes towards ethnic minorities among the native majority.

The connection between preferences for homogeneity and actual segregation

It should be assumed that countries with higher preferences for ethnic homogeneous neighbourhoods also would have stronger ethnic segregation and concentration. This does not seem quite to be the case when looking at Figure 8.11.

In the figure we have ordered some countries according to the strength of their preferences for homogeneous neighbourhoods and have for the same countries

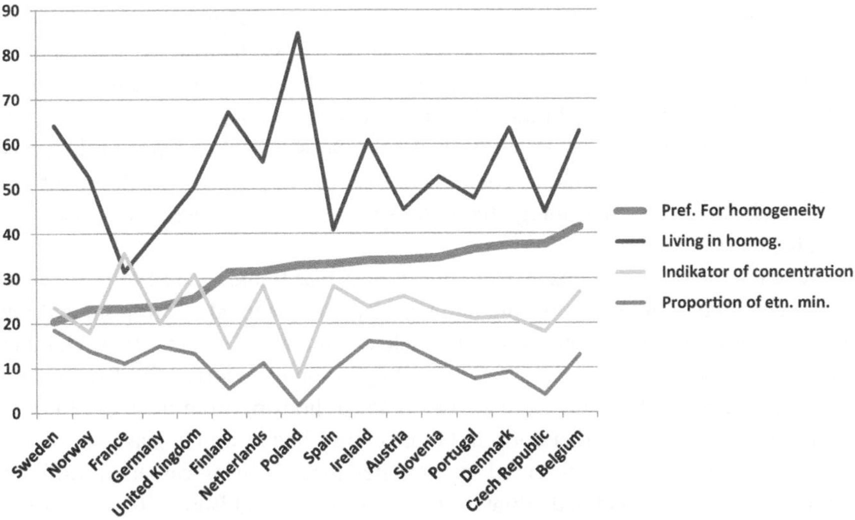

Figure 8.11 Countries ordered according to preferences for ethnic homogeneous neighbourhoods (2002), showing also the proportion living in homogeneous neighbourhoods, the calculated indicator of ethnic concentration and proportion of foreign-born residents (2015) in the countries

Sources: Semyonov and Glikman 2009, European Social survey (ESS) 2002 and United Nations Department of Economic and Social Affairs 2015 report: Trends in International Migrant Stock

shown the previously calculated indicator of ethnic concentration. A systematic increasing indicator of concentration with increasing preferences for homogeneity is not found across countries. There are some fluctuations but not an increasing tendency. There is also not a straightforward connection between preferences for homogeneity and the proportion of residents actually living in such neighbourhoods.

The figure also shows that there is not a systematic connection between preferences for homogeneity and the proportion of ethnic minorities in the countries, measured as previously in Figure 8.3 as the proportion of foreign born.

The somewhat curious conclusion from this is that even if it is demonstrated that negative attitudes towards ethnic minorities affect preferences for not mixing with ethnic minorities, it does not seem to have consequences for the extent to which the actual residence of the native majority is separated from ethnic minorities. The explanation might be that differences across countries concerning residential preferences among ethnic minorities and the native majority do not constitute a decisive determination of ethnic segregation. As described in Chapters 6 and 7, ethnic segregation depends very much on the residential options for both ethnic minorities and natives constituted by housing markets and structural properties of cities.

Housing policies and markets across countries

As shown in Figure 8.3, the extent and character of immigration have diverged greatly across European countries. This is partly due to the historic development where some countries have received an early immigration from their former colonies (Britain, Netherlands, France, Belgium, Spain and Portugal) while others had their first immigrants as guest workers in the 1960s. This has affected the housing situation of immigrants, as some have been in the country much longer and have had time to become integrated and establish housing careers (as discussed in Chapter 4).

When the first immigrants arrived after World War II and in the 1960s, they had very restricted options for housing and were mostly confined to informal and low quality housing in the central cities (Özüekren and Van Kempen 2003; Hamnett and Butler 2010; Drever and Clark 2002). Only in some countries, like Sweden and Denmark, did they have the right to live in social housing from the start. In other countries, such as the Netherlands and France, they had to wait at least a decade to obtain housing in the social rented sector. After 1980, housing careers of different immigrant groups began to diverge in Britain (Hamnett and Butler 2010) and other countries, where some moved into social housing and others to homeownership in the suburbs. In other countries like Austria, ethnic minorities have until recently been excluded from certain segments of the subsidised rent sector. In countries where social rented housing does not exist or where access to it has been restricted, minority ethnic groups had to solve their housing problems within the limitations of the private sector. For example, in Germany special difficulties have been faced by minority ethnic groups in the private housing sector (Drever and Clark 2002).

In some countries ethnic minorities still have to opt for low quality and overcrowded old housing. This is especially the case in Southern-European countries (Malheiros 2002) where informal housing markets have provided opportunities for immigrants both in the city centres and outside. Parts of this housing are owner-occupied. Homeownership among ethnic minorities has also been growing in other countries, mostly because of increasing integration and economic improvements, but also in some countries because of lack of alternatives because social housing is restricted and private renting difficult to access (e.g. Norway). In countries with easy access to social housing, like Denmark, Sweden, Finland, France and the Netherlands, a considerable proportion of ethnic minorities have settled in social housing. General statistics about this have not been found, but as is demonstrated in Chapter 6, the proportion of ethnic minorities living in social housing is 45 per cent in Denmark, 43 per cent in Finland and 20 per cent in Sweden. Whitehead and Scanlon (eds.) (2007) reported that the figures at that time were 51 per cent in the Netherlands, 30 per cent in France, 28 per cent in England and more than 20 per cent in Austria.

Housing policy models

In the research literature a distinction has been made between 'dual' and 'unitary' housing markets. In dual markets homeownership is dominated by upper income groups, while families with lower incomes are more confined to the rental market (see Chapter 6). Dual markets are also called segmented markets, because low-income households are more concentrated in rental housing while higher income groups tend to have obtained homeownership. This to a greater extent limits the options of ethnic minorities to rental housing. Another distinction is to divide between unitary and dual rental markets. In the dual markets social housing is a smaller or much restricted sector for the poorest families. In some of the research literature a hypothesis has been put forward that especially these kinds of markets will result in a high concentration of ethnic minorities in social housing.

As described in Chapter 6, housing policies vary immensely across countries, and for different reasons there is not always a straightforward connection between housing policies and welfare policies in general. Some countries with strong welfare have chosen more liberal housing models, while some countries with weak welfare have implemented a more social housing policy, as it will appear from what follows.

Tenure composition

Figure 8.12 shows an estimate of tenure composition of housing markets in different European countries.

It appears from the figure that countries in the Latin Rim cluster and in Eastern Europe nearly all have a low proportion of rented housing (except Czechia and Poland) and a small stock of social housing. Some of the southern countries also have up to 20 per cent private renting. The largest non-ownership sectors are found in the Social Democratic and Corporatist clusters. Social housing is also mostly found in Social Democratic countries, but is also important in some other countries like Austria, France and the UK. Social housing is nearly non-existing in many countries, especially in Southern and Eastern Europe.

According to Arbaci (2007), the Social Democratic cluster mostly has unitary rental housing systems. This does not apply, however, for Finland and Norway, as shown in Chapter 6. The Corporatist welfare cluster is expected to have unitary rental systems. France and Austria have relatively large social housing sectors, but Germany and Belgium do not. Some countries in the Liberal cluster, the UK and Switzerland, also have a considerable social housing sector. The extent to which rental housing is dual or unitary depends, however, on how social housing is organised.

Social housing models

The options for ethnic minorities in the housing market and the difficulties they meet in their efforts to get decent housing are very much dependent on the supply of housing in different housing tenures and how housing policy regulates access and housing costs. The supply of low cost rental housing, especially social housing, and the rules for access to this housing, are very important for ethnic minorities.

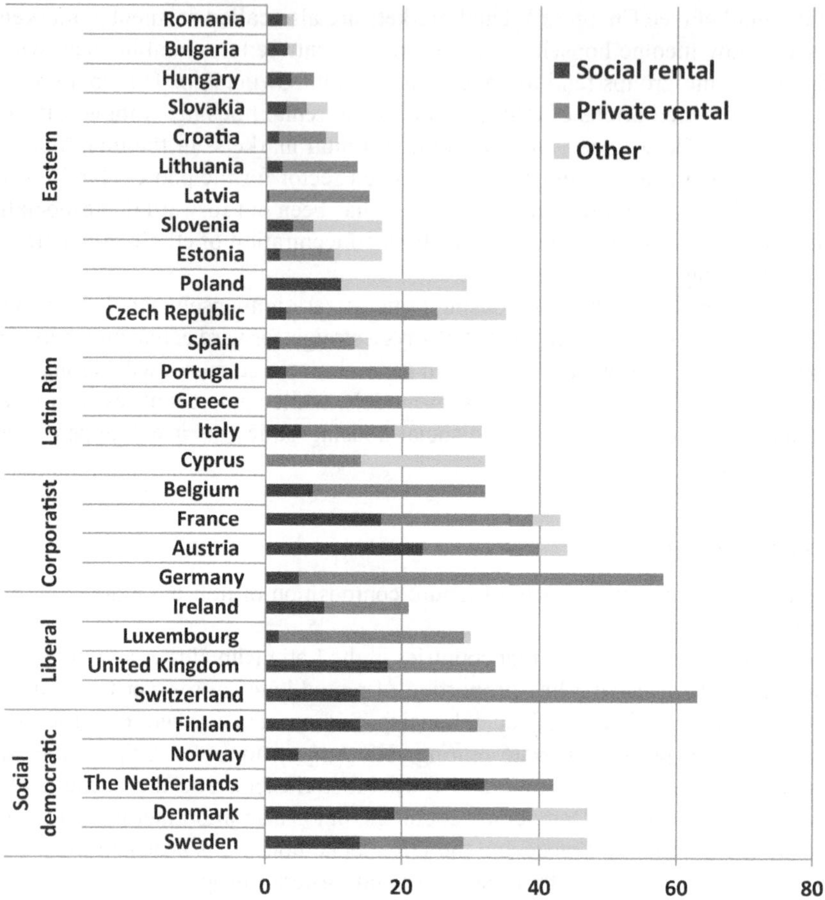

Figure 8.12 Proportion (%) of dwellings in the housing stock that are not owner-occupied housing, divided by tenures

Source: Rosenfeld 2015

The supply of social housing and rules for access vary greatly across countries. Sometimes social housing is open for all house hunters and sometimes access is restricted to families with low incomes or specific needs. There is usually a connection between extent of supply and rules for access where countries with small social housing sectors also have restrictions on access. Large social housing sectors with free access are the most favourable for ethnic minorities. Social housing systems across Europe have been divided into three different models (Rosenfeld 2015).

The *Universal allocation model* was designed to make decent and affordable housing available to the whole population regardless of the income. In this model, housing is considered a public responsibility and is provided either through municipal housing companies (e.g. Sweden) or through non-profit organisations. The *Generalist allocation model* assumes that the objectives of housing policy will be predominantly met by the market. Social housing is aimed only at those households for which the market is unable to deliver housing of decent quality at an affordable price. The *Residual allocation model* aims to allocate housing to a restricted category of beneficiaries, usually very vulnerable households that are heavily dependent on state benefits (e.g. unemployed, disabled, elderly, single parents).

According to Rosenfeld (2015), only 3 EU countries can be seen as Universal models (Denmark, Sweden and the Netherlands), 13 have used the Generalist models (Austria, Czechia, France, Finland, Belgium, Germany, Poland, Slovenia, Italy, Luxemburg, Greece, Spain and Slovakia). Some of these countries (France, Finland, Belgium, Germany and Slovakia) have also used the Residual model in some of the social housing. Other countries (the UK, Estonia, Ireland, Hungary, Cyprus, Portugal, Bulgaria, Lithuania, Latvia, Romania and Estonia) have only used the Residual model. However, recent changes in policies have changed the position of countries and the social housing stock is increasingly moving towards the Residual social housing provision. This study indicates that the Universal models mostly are used in Social Democratic welfare states, the Generalist models mostly in Corporatist and Latin Rim countries and the Residual models in the Liberal welfare and Eastern clusters. Some countries do not, however, fit into this picture, as some of the Eastern and Latin Rim countries use a Generalist model.

Social segmentation of housing markets

As discussed in Chapter 6, the housing options of ethnic minorities are influenced by the options for low-income groups in general, because they mostly have lower incomes. Ethnic minorities are, as shown in Chapter 2, often dependent on the availability of rented housing, especially in housing markets where low-income groups are confined to this tenure. The extent to which there is a social division, called segmentation, of housing markets affects the housing options of ethnic minorities.

It is possible to illustrate the degree of *income segmentation* regarding home-ownership/renting in some European countries based on data from the *First European Quality of Life Survey: Social dimensions of housing* conducted by the European Foundation for the Improvement of Living and Working Conditions (2006). The survey provides data on homeownership rates in four income quartiles in each country from which an index of homeownership segmentation can be calculated, which is seen in Figure 8.13.

It is seen that income segmentation of housing markets varies greatly across countries, also within welfare clusters. There is not a straightforward connection between welfare state cluster and dual/unitary housing markets. The five countries with the highest income segmentation are spread across all four clusters. The most important factor seems to be that countries with high segmentation often also have

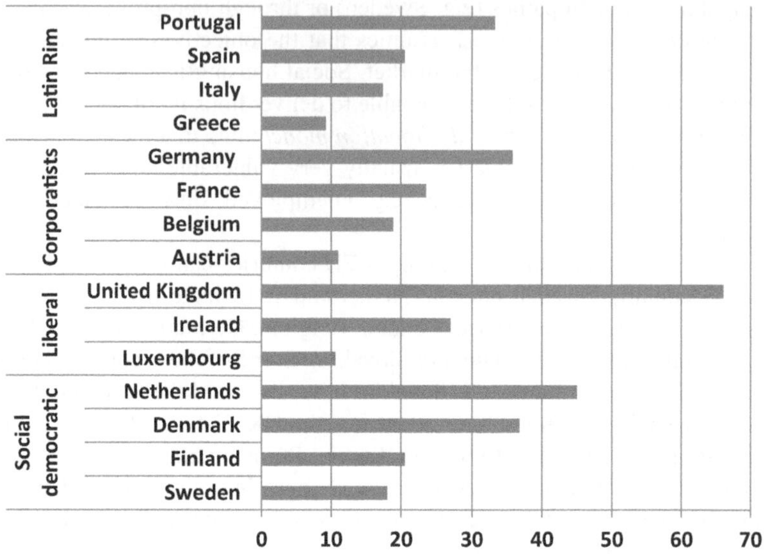

Figure 8.13 Calculated indices of income segmentation for homeownership/renting in housing markets

Note: The calculations are based on data from European Foundation for the Improvement of Living and Working Conditions (2006) on proportion of homeownership in four income quartiles. Index of segmentation = $\text{sum}_{x=1,4}$(abs(proportion of homeowners in quartile x – proportion of homeowners for the whole population))/ proportion of homeowners for the whole population *100

larger rental sectors (see Figure 8.12). This is, however, not universal, as Sweden, for example, has weak segmentation, and the UK with strong segmentation only has a moderate proportion of rental housing. This indicates that lower segmentation in some countries might be a result of families with lower incomes being forced into homeownership, often of low quality, because of lack of rental housing. This can, as is indicated in the study from the Nordic countries referenced in the last section of Chapter 6, have markedly negative consequences for the housing situation for ethnic minorities.

Most of the countries in the Social Democratic cluster should have unitary housing markets, but segmentation is quite high in the Netherlands and Denmark. Three of them, the Netherlands, Denmark and Sweden, have unitary rental markets, while Finland does not (see Chapter 6). Norway, which is also part of the cluster, is an exception with a very small and restricted social housing sector (Skifter Andersen 2012). Also in Finland social housing is a somewhat restricted sector, but somewhat larger than in Norway.

Countries in the second cluster have, except for Austria, a dual rental market and, in general, a certain rental sector. In two of the countries, France and Austria, there is a larger social housing sector. Belgium, however, more closely resembles the third group, the Liberal welfare cluster, which has a smaller rental sector and a strongly restricted social housing sector. Especially in the UK there is a very dual housing market.

Finally, the Latin Rim cluster is heavily dominated by homeownership with a small and very restricted social housing sector. Income segmentation differs greatly, from Portugal with a high segmentation to Greece with a more equal distribution of homeownership. According to Arbaci (2007 and 2008), the Latin Rim countries highly rely on informal family and voluntary agencies in the provision of welfare services, including housing. As will be discussed subsequently, these differences in housing systems across countries only to some extent affect ethnic segmentation of housing markets.

The housing situation of ethnic minorities

Differences in homeownership rates between ethnic minorities and the native majority

Figure 8.14 shows the homeownership rate for residents in different EU countries who do not have citizenship in any EU country. In the left part of the figure this is compared with homeownership for the whole population. The data thus does

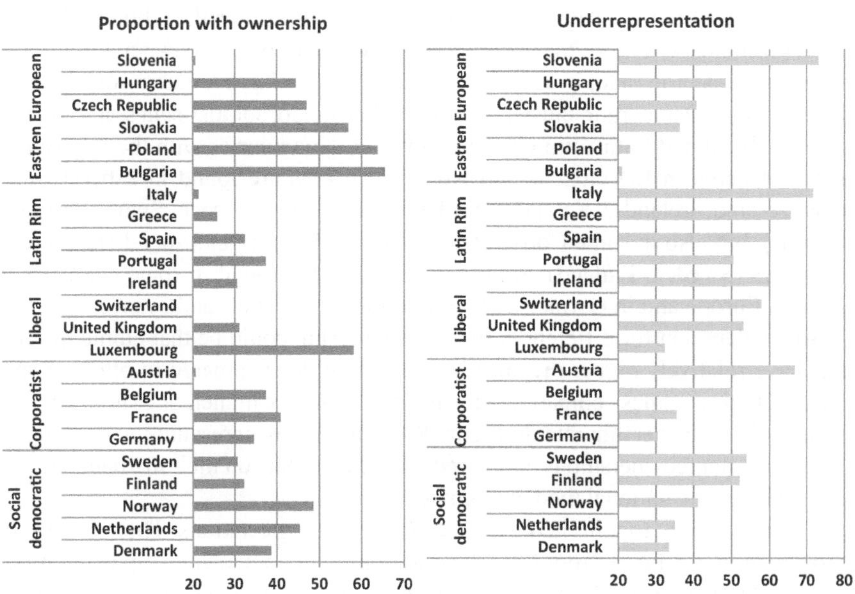

Figure 8.14 Proportion (%) of Non-EU citizens, 18–64 years old, who are living in homeownership 2016, and *ethnic segmentation* measured as the extent to which they are under-represented (%) in owner-occupation compared with the whole population (EU citizens, including Norwegians and Swiss)

Source: EUROSTAT

not encompass all ethnic minorities, only those who have not obtained citizenship in EU. It can be expected that the included persons are less integrated and have a more vulnerable housing situation than average. On the other hand immigrants from other Western countries are also included who more often are living in homeownership (see Chapter 2).

The data is highly influenced by the kind of immigration that has occurred in the different countries. In Eastern Europe there has only been little immigration and most of the immigrants are coming from other Non-EU countries in Eastern Europe. Like natives, a large proportion of ethnic minorities have homeownership. In all Western European countries, except for Luxembourg, less than half of the Non-EU citizens have obtained homeownership, and there does not seem to be a significant connection between welfare cluster and homeownership. Countries with the fewest homeowners are found in Southern and Central Europe (Switzerland, Austria, Slovenia, Italy and Greece), followed by some countries from Northern and Central Europe, but the pattern is not clear.

The explanation for the differences could be that homeownership in general varies across countries. The right part of the figure shows, however, that this is not the case. The extent of homeownership among Non-EU citizens is here calculated as the relative difference in homeownership rates compared to the whole population of EU citizens in the countries. This data sheds light on what can be called the *ethnic segmentation* of housing markets; the extent to which ethnic minorities are separated from natives in different housing tenures. For the whole EU, Non-EU citizens have a more than 50 per cent lower rate of homeownership. Ethnic segmentation varies a lot inside welfare clusters, but it has a tendency to be lowest in the Social Democratic cluster, a little higher in Corporatist states and highest in the Liberal and Latin Rim clusters. Some of the countries with the largest differences between EU citizens and Non-EU citizens are found in Liberal states (Switzerland, Ireland, UK) and Latin Rim States (Italy, Greece, Spain, Portugal, Cyprus), but also in some Corporatist countries (Austria) and Social Democratic countries (Sweden, Finland). Smaller differences are found in some countries in Eastern Europe and in Germany, Denmark, the Netherlands and France.

An explanation for differences in homeownership could be that ethnic minorities have relatively higher incomes in some countries compared to others. Especially if their incomes compared to the national level are higher, one should expect a higher homeownership rate. In Figure 8.15, the under-representation of Non-EU citizens in homeownership is compared with the proportion their median incomes constitute of the national median income.

The figure only to a limited extent confirms the expectations. There is a slight movement in the figure from lower representation in homeownership in countries where ethnic minorities' incomes are lower compared to the national level, to relatively higher homeownership rates in countries where their incomes are relatively higher. However, the picture is quite diffuse.

This data insinuates that there is not a simple connection between welfare state systems, income levels and ethnic segmentation of housing markets. The extent and character of immigration might have had some influence on the differences

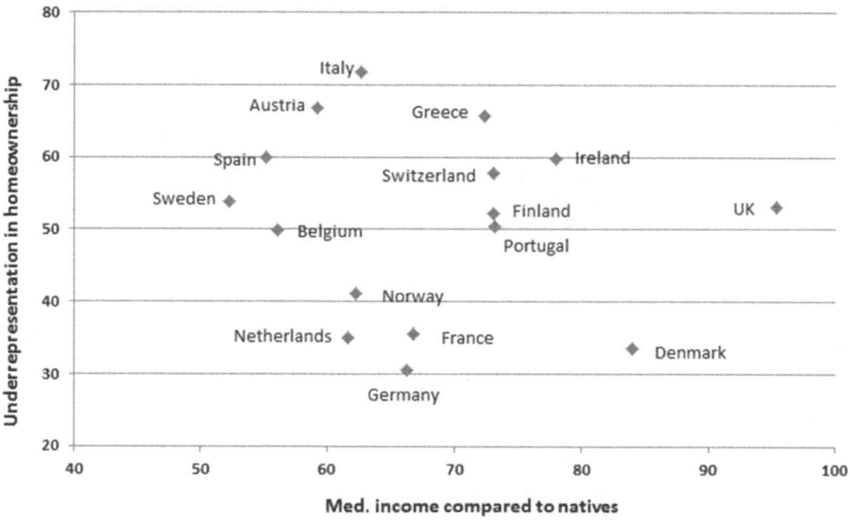

Figure 8.15 Comparison of the under-representation (%) of Non-EU citizens in homeownership in countries with the proportion (%) their median incomes represent of the national median income

Source: EUROSTAT

as some countries have received more immigrants with low education or refugees with social problems. Most importance must, however, be attached to differences in housing policies and housing markets across welfare clusters that provide different opportunities for low-income groups and for ethnic minorities. It should be assumed that if homeownership is more difficult to access for low-income groups, it will also be more difficult for ethnic minorities because they in general have lower incomes.

In Figure 8.16, under-representation of ethnic minorities in homeownership is compared with the previously calculated income segmentation for homeownership/renting in some Western European countries. It appears from the figure that there is not a simple connection between income segmentation and this measure of ethnic segmentation, but that there seems to be a reverse relationship between income and ethnic segmentation. The three countries with the highest ethnic segmentation (Italy, Austria and Greece) also have quite low income segmentation. The UK has the highest income segmentation, but only a mediate ethnic segmentation. The Figure again points towards that ethnic segmentation of housing markets cannot be explained simply by welfare state systems and the typology of dual/unitary housing markets, as discussed in Chapter 6. Besides differences in immigration to the countries the reason for this might be that the character and size of the rental markets, which differs much across welfare clusters and market types, has decisive importance.

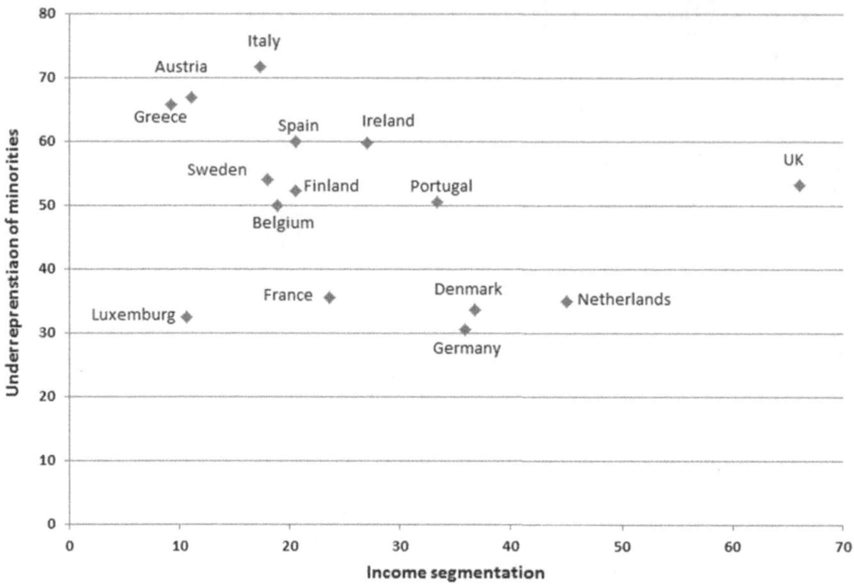

Figure 8.16 Comparison of under-representation of ethnic minorities in homeownership with income segmentation of housing markets in Western European countries

Note: Calculated as in Figure 8.13

Comparative data on the connection between housing market conditions and ethnic segregation across European countries is difficult to obtain. It is possible, however, to use data from EUROSTAT and the European Social Survey to produce figures that to some extent are indicators on the connection between ethnic segmentation of housing markets and ethnic segregation in the countries. Figure 8.17 shows the connection between segmentation, calculated as under-representation of ethnic minorities in homeownership, and proportion of residents in the countries who in the ESS answered that they lived in a neighbourhood *with many minorities*. The last figures are not completely reliable measures of ethnic segregation, as they depend on the level of immigration to the countries, such that in countries having a high proportion of ethnic minorities the likelihood of living in a multi-ethnic neighbourhood might also be higher.

Nevertheless the figure indicates that living in multi-ethnic neighbourhoods is more frequent in countries with a high segmentation of housing markets. France is a deviant case, but this cannot be explained by a higher representation of ethnic minorities in the country, as is seen from Figure 8.3.

Discrimination in the housing market

It has been shown in the preceding, and in Chapter 6, that the housing situation of ethnic minorities cannot be explained by their lower incomes. Other

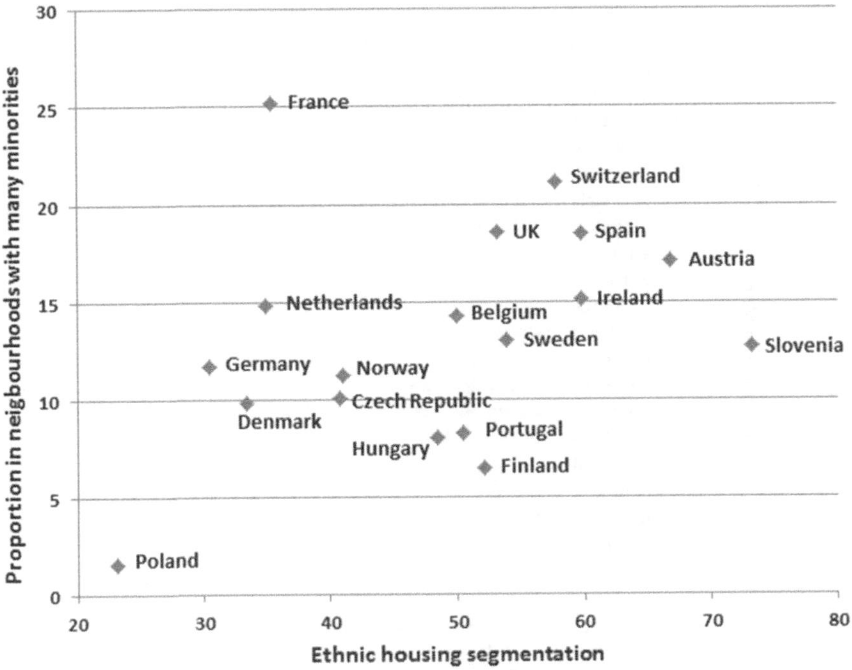

Figure 8.17 The relationship between ethnic housing segmentation of housing markets in European countries (measured as under-representation of ethnic minorities in homeownership (%)) and the proportion of residents (%) stating that they live in multi-ethnic neighbourhoods

Sources: EUROSTAT and European Social Survey 2014

explanations are important, as described in Chapter 6. Among the important conditions for housing options of ethnic minorities is discrimination in the housing market. Data from Second European Union Minorities and Discrimination Survey made by the European Union Agency for Fundamental Rights (2017) can be used to illuminate the extent of discrimination in the housing sector for different ethnic groups in different countries. Figure 8.18 shows for four different ethnic groups their perception of experienced discrimination in housing in different countries.

It is clear that discrimination varies greatly across ethnic groups and countries. On average North Africans seem to be the most discriminated against group, followed by Sub Saharan Africans and Turks, but the differences can to some extent be explained by differences across the countries where they live. Discrimination is mostly found in Latin Rim countries, while it is lowest in the Social Democratic welfare states and in the UK. The Corporatist states, Belgian, Germany, Austria and France, all have quite high degree of discrimination.

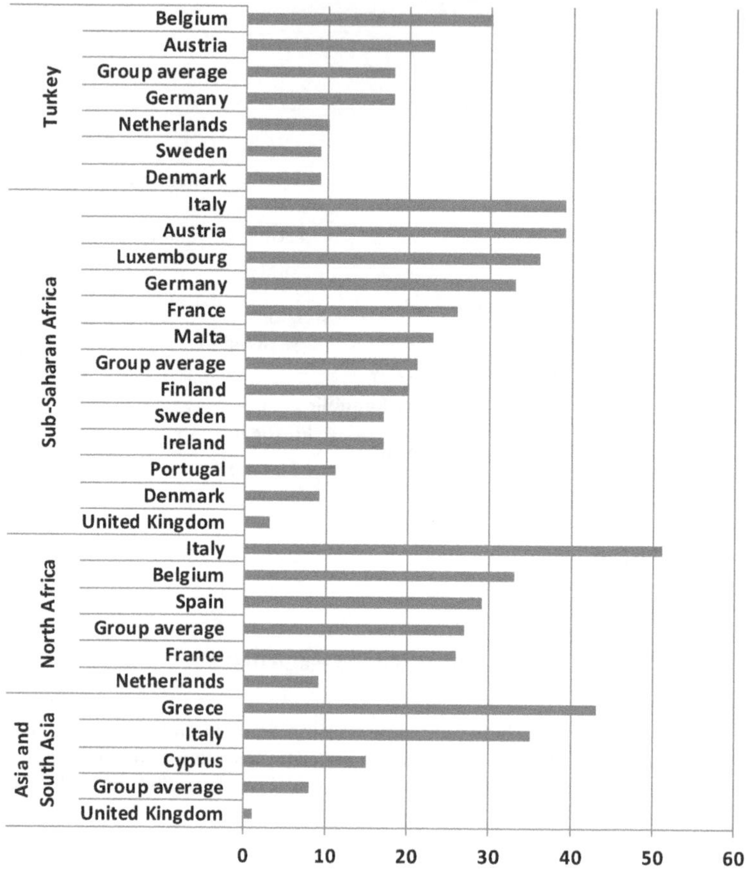

Figure 8.18 Proportion of respondents of selected ethnic groups in selected countries, who said yes to the question: In the past 5 years have you ever felt discriminated against because of skin colour/ethnic origin/religion when trying to rent or buy housing?

Source: Data from the Second European Union Minorities and Discrimination Survey 2017, European Union Agency for Fundamental Rights (2017)

Ethnic minorities in overcrowded housing

EUROSTAT has also provided some figures for the degree of overcrowding among Non-EU citizens, as is shown in Figure 8.19.

The figures should be handled with care, as there are different definitions across countries of the kinds of rooms that are included. In some countries kitchen and bath rooms are incorporated, while others only count sitting and bedrooms, and some countries do not have reliable population statistics. Especially the figure

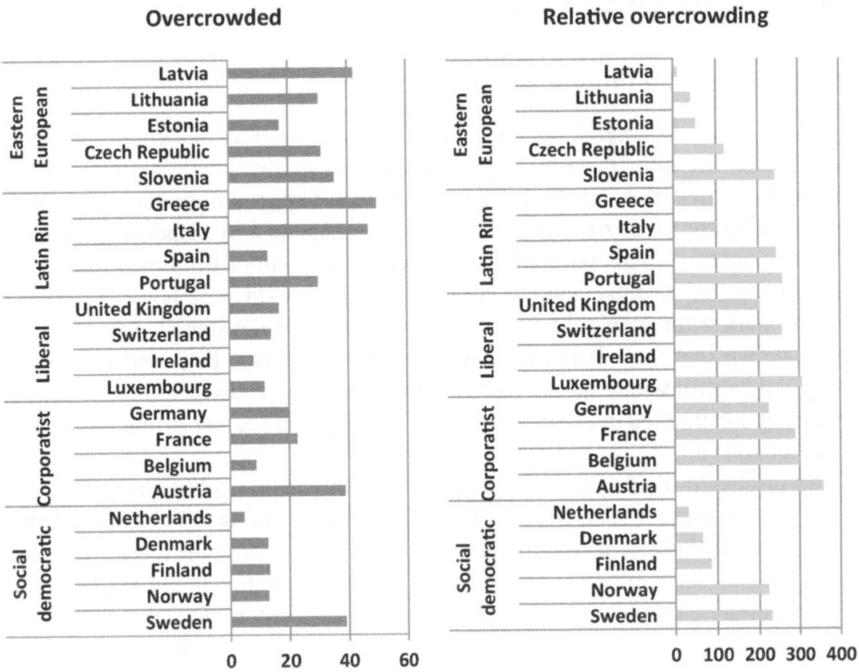

Figure 8.19 The proportion (%) of Non-EU citizens, 18–64 years old, who were living in overcrowded housing in 2016, and the relative differences in their overcrowding (%) compared to EU citizens

Source: EUROSTAT

Note: A person is in this statistic considered as living in an overcrowded household if the household does not have at its disposal a minimum number of rooms equal to: one room for the household; one room per couple in the household; one room for each single person aged 18 or more; one room per pair of single people of the same gender between 12 and 17 years of age; one room for each single person between 12 and 17 years of age and not included in the previous category; one room per pair of children under 12 years of age.

from Sweden seems to be much too high compared to the figure shown in the last part of Chapter 6, while the figures from Norway and Finland are too low.

Overcrowding among ethnic minorities is especially often found in Latin Rim and Eastern countries. In the right part of the figure is calculated the relative differences between Non-EU and EU citizens, which is a more reliable figure as the data for natives and ethnic minorities are handled in the same way. The average for the EU is that Non-EU citizens about 80 per cent more often live in overcrowded housing, as measured in accordance with the EUROSTAT definition. The largest relative *housing deprivation* among Non-EU residents seems most often to be found in countries belonging to the Liberal and the Corporatist countries, the smallest in Eastern Europe and some countries in the Social Democratic cluster.

Again the differences across countries might be assigned to different kinds of immigration (Eastern Europe) and housing markets (Social Democratic cluster).

Planning and building systems

As discussed in Chapter 7, both the spatial structure of cities and ethnic and social segregation are influenced by the way public authorities intervene in the development of cities by controlling the use of land. Of special importance is to what extent different kinds of housing are spatially dispersed. A dispersal of housing for low-income groups will only take place if local authorities through planning and steering of land use make it possible to build low-income housing in places where market forces would result in more expensive housing. It has, however, been shown that there is not a straightforward connection between strong planning and control with land use and mixing of housing tenures. Sometimes strong planning has resulted in high separation of housing tenures and strong segregation (Arbaci 2007).

As discussed in Chapter 7, it emerges from the theoretical planning literature (Nadin and Stead 2008; Dasi 2007) that spatial planning systems are deeply embedded in their socio-economic, political and cultural context and that planning systems seem to be connected to welfare state systems. In the Social Democratic welfare cluster of countries, and also in some Corporatist welfare states like Germany and Austria, planning systems where public authorities are intervening heavily in the development of cities have been common. In countries mostly found in the Liberal welfare cluster there has been used a much narrower kind of regulation with the limited purpose of regulating land use change and with an urban development created mostly by market forces. In other countries, from the Corporatist and Latin Rim clusters, planning has mostly been steered by regional objectives of economic growth and development, where local objectives of developing cities with less spatial inequality have not been much in focus. Finally, most Latin Rim countries have had a spatial planning style which takes place on the local level through building regulations. Public planning is weak and seldom interferes with the use of land.

In her comparison of ethnic segregation and its causes in Western European countries, Arbaci (2007) showed systematic differences across these groups of countries. Based on her data, collected from different sources on segregation of single ethnic groups in different cities, she concluded that the Nordic Social Democratic welfare states (and the Netherlands), despite lower housing segmentation, have a relatively high degree of ethnic segregation (although somewhat lower than that of the Liberal welfare states). She explains this contradiction by the mode of building production, where building activities in Social Democratic welfare states are dominated by large-scale builders and developers. She argues that in spite of strong planning systems and much public land ownership, the structure of the building and housing sector with large builders, developers and owners has promoted the development of large urban areas with homogeneous tenures and building types, which has promoted a spatial segregation of tenures. She also found that Liberal welfare regimes tend to have large builders/developers relying more on speculative development gains than on building profits. Given the size of developers, a considerable large scale of production and of spatial differentiation should be expected in both clusters.

The Corporatist welfare state cluster was characterised by having a more fragmented building industry than in both former cases. The Latin Rim welfare states of the Mediterranean should have even smaller builders than their Corporatist counterparts. In both Corporatist and Latin Rim regimes, a division is found between different kinds of firms that perform housing production for respectively high, middle and low-income households, and are producing at a relatively small scale (Leal 2004).

Another conclusion from Arbaci's (2007) study was that until the early/mid-1990s, in Corporatist and Social Democratic clusters, public provision of land had provided the arena for a more affordable and accessible land supply. This has been essential, particularly for self-provision, cooperatives and other non-profit housing forms, as well as for the large-scale post-war urban expansion – counter-urbanisation of middle and high-income households. At the other extreme, among most Latin Rim welfare states, the patrimonial tradition in land ownership and the scarcity of affordable developable land have, by necessity and constraints, developed alternative forms of land supply accounting for substantial informal land-property markets and informal housing provision, for both homeownership and rent (Leal 2004). This provided the uneven, small-scale and fragmented pattern of urban sprawl – suburbanisation of low and middle-low-income households.

According to Arbaci, we should expect diverse types and levels of spatial segregation across the welfare clusters; higher in Liberal and Social Democratic welfare states, more fragmented in Latin Rim welfare states and lower in Corporatist welfare states. Her analyses were, however, based on data on the segregation of single ethnic groups, while we have used data on the segregation of the whole group of ethnic minorities. The results shown previously in Figures 8.5 and 8.6 are not quite in accordance with the evidence found by Arbaci. Corporatist welfare states seem to have a relatively high concentration of ethnic minorities, especially in France, followed by the UK and Germany. Scandinavian countries and to some extent Holland were found to have a lower level of ethnic concentration. Southern countries like Spain and Greece were measured to have medium segregation. The points made by Arbaci can, however, partly illuminate why countries in the Social Democratic and Corporatist clusters still have relatively strong segregation despite less social inequality, more rented housing, less segmentation of housing markets and stronger public planning.

Do welfare systems affect ethnic segregation?

The subject of this chapter has been to examine if some of the differences in ethnic segregation across European countries can be explained by differences in their societal and economic organisation expressed in so-called welfare state systems. We have used the typology defined in earlier research about such systems, dividing the countries into five clusters that to a different content and extent have provided welfare services for their citizens.

The data used on to what extent minorities and natives think that they live in multi-ethnic neighbourhoods suggests that there are some differences in ethnic segregation across different welfare clusters. However, it does not indicate a straightforward connection between the extent of welfare and segregation. Segregation seems to be the lowest in the most extensive welfare states, the Social Democratic cluster, but it is also quite low in some of the countries with weaker welfare systems in Southern Europe. The highest segregation has been found in the Corporatist cluster, with medium strong welfare systems, and in the Liberal cluster, with weaker systems.

A possible explanation for the differences might be variation in social inequality across countries which affects the economic situation of ethnic minorities. Income inequality is higher in the countries with the least extensive welfare systems, but this does not seem to affect the incomes of ethnic minorities in comparison with the native majority in countries. The relationship between the median income of ethnic minorities and the national average does not vary much across welfare clusters. Moreover, a comparison of the Gini index for the countries with the indicator of ethnic concentration does not show a systematic connection between these two measures. Income inequality thus does not automatically lead to ethnic segregation, and differences in social inequality across countries thus cannot explain variation in ethnic segregation.

It is demonstrated that spatial concentration of ethnic minorities to some extent is connected to the size of the ethnic minority population in the countries. As expected, the presence of neighbourhoods with many ethnic minorities will, everything else being equal, be more common in countries with a larger minority population. Moreover, it is possible that processes of segregation, as discussed in Chapters 3 and 5, might be stronger in cities with more ethnic minorities. The possibilities of forming ethnic enclaves that make it more attractive for ethnic minorities to move into concentration areas are better, and white flight and avoidance might be stronger, leading to stronger preferences for living in neighbourhoods without ethnic minorities.

Segregation is influenced by the preferences for location among ethnic minorities and the native majority. Preferences for living in ethnic homogeneous neighbourhoods vary greatly across countries and also inside welfare clusters, but it is indicated that they in general are not stronger in countries with many ethnic minorities than in countries with fewer. These preferences are stronger in countries where the population shows more negative attitudes towards ethnic minorities, and such attitudes are more often found in countries with less extensive welfare systems, but this does not seem to have consequences for the extent to which the actual residence of the native majority is separated from ethnic minorities. The explanation might be that differences concerning residential preferences among ethnic minorities and the native majority across countries do not constitute a decisive determination of ethnic segregation. As described in Chapters 6 and 7, ethnic segregation depends very much on the residential options for both ethnic minorities and natives constituted by housing markets and structural properties of cities.

The housing situation of ethnic minorities, measured by living in overcrowded housing, is better in the more developed welfare states than in Southern and Eastern Europe. This is a result of differences in housing policies and housing markets and points towards that the housing options of ethnic minorities are better in developed welfare states. However, in comparison with the general housing situation in the countries, ethnic minorities do worst in the Corporatist and the Liberal cluster.

Housing policies vary greatly across countries, and for different reasons there is not always a straightforward connection between housing policies and welfare policies in general. Some countries with strong welfare have chosen more liberal housing models, while some countries with weak welfare have implemented a more social housing policy. Countries in the Latin Rim cluster and in Eastern Europe nearly all have a low proportion of rented housing and a small stock of social housing, but some of the southern countries have some private renting. The largest non-ownership sectors are found in the Social Democratic and Corporatist clusters. Social housing is also mostly found in Social Democratic countries, but is also important in some other countries like Austria, France and the UK. Social housing is nearly non-existent in many countries, especially in Southern and Eastern Europe.

The comparison of countries has shown that there seems to be a connection between the degree of ethnic spatial concentration and the extent to which ethnic minorities are separated from natives in rental and owned housing, so-called ethnic segmentation of the markets. In countries where ethnic minorities are more confined to rental housing, ethnic spatial concentration is higher. Ethnic segmentation varies greatly inside welfare clusters, but it tends to be lowest in the Social Democratic cluster, a little higher in Corporatist states and highest in the Liberal and Latin Rim clusters. This variation cannot be explained by differences across countries in welfare indicators such as income levels and income inequality.

There is also no connection between the degree of income segmentation of housing markets and ethnic segmentation, indicating that ethnic minorities' access to homeownership is less dependent on to what extent housing policies create dual housing markets, but has other determinants. Important is if minorities have easy access to the rental sector, where discrimination plays a role. Discrimination is mostly found in Latin Rim countries, while it is lowest in the Social Democratic Welfare states and in the UK. The Corporatist states, Belgian, Germany, Austria and France, all have quite high degree of discrimination. Most important is access to social housing, determined by the size and regulation of the sector.

In some countries in the Social Democratic and Corporatist clusters, ethnic minorities are to a high degree concentrated in social housing and the spatial location of this housing is of basic importance for ethnic segregation. More developed planning systems in these countries should in theory have made it possible to spread out social housing in the city, but for different reasons these have not always been used, leading to large neighbourhoods with concentrations of social housing and ethnic minorities. Sometimes this development has been a result of a structure of the building and housing sector with large builders, developers and owners.

In the Liberal and Latin Rim countries, ethnic minorities are more dispersed in the housing stock, and the location of ethnic minorities is more dependent on market processes of social and ethnic segregation. Minorities have often been confined to low quality housing in private renting or homeownership concentrated in more or less declined neighbourhoods.

References

Arbaci S. (2007) Ethnic Segregation. Housing Systems and Welfare Regimes in Europe, *International Journal of Housing Policy*, 7(4), pp. 401–433.

Arbaci S. (2008) (Re)Viewing Ethnic Residential Segregation in Southern European Cities: Housing and Urban Regimes as Mechanisms of Marginalisation, *Housing Studies*, 23(4), pp. 589–613.

Dasi F. J. (2007) *Governance of Territorial and Urban Policies from EU to Local Level.* Final Report of ESPON Project 2.3.2. ESPON Coordination Unit, Esch-sur-Alzette.

Drever A. and Clark W. A. V. (2002) Gaining Access to Housing in Germany: The Foreign Minority Experience, *Urban Studies*, 39, pp. 2439–2454.

Esping Andersen G. (1990) *The Three Worlds of Welfare Capitalism.* Cambridge: Polity Press.

European Foundation for the Improvement of Living and Working Conditions (2006) *First European Quality of Life Survey: Social Dimensions of Housing.* Dublin.

European Union Agency for Fundamental Rights (2017) *Second European Union Minorities and Discrimination Survey.* Vienna.

Glikman A. and Semyanov M. (2012) Ethnic Origin and Residential Attainment of Immigrants in European Countries, *City & Community*, 11(2), pp. 198–219.

Hamnett C. and Butler T. (2010) The Changing Ethnic Structure of Housing Tenures in London, 1991–2001, *Urban Studies*, 47(1), pp. 55–74.

Iceland J., Mateos P. and Sharp G. (2011) Ethnic Residential Segregation by Nativity in Great Britain and United States, *Journal of Urban Affairs*, 33(4), pp. 409–429.

Johnston R., Forrest J. and Poulsen M. (2002) Are There Ethnic Enclaves in English Cities? *Urban Studies*, 39, pp. 591–618.

Leal J. (2004) Are Southern Housing Systems Different? In Allen J. et al. *Housing and Welfare in Southern Europe*. Oxford: Blackwell.

Malheiros J. M. (2002) Ethni-Cities: Residential Patterns in Northern European and Mediterranean Metropolis. Implication in Policy Design, *International Journal of Population Geography*, 8(2), pp. 107–134.

Musterd S. (2005) Social and Ethnic Segregation in Europe: Levels, Causes and Effects, *Journal of Urban Affairs*, 27(3), pp. 331–348.

Musterd S. and Van Kempen R. (2009) Segregation and Housing of Minority Ethnic Groups in Western European Cities, *Tijdschrift voor Economische en Sociale Geografie*, 100(4), pp. 559–566.

Nadin V. and Stead D. (2008) European Spatial Planning Systems, Social Models and Learning, *The Planning Review*, 44(172), pp. 35–47.

Özüekren A. S. and Van Kempen R. (2003) Special Issue Editors' Introduction: Dynamics and Diversity: Housing Careers and Segregation of Minority Ethnic Groups, *Housing, Theory and Society*, 20(4), pp. 162–171.

Rosenfeld O. (2015) *Social Housing in the UNECE Region. Models, Trends and Challenges*. Geneva: United Nations Economic Commission for Europe.

Semyonov M. and Glikman A. (2009) Ethnic Residential Segregation, Social Contacts, and Anti-Minority Attitudes in European Societies, *European Sociological Review*, 25(6), pp. 693–708.

Semyonov M., Glikman A. and Krysan M. (2007) Europeans' Preference for Ethnic Residential Homogeneity: Cross-National Analysis of Response to Neighborhood Ethnic Composition, *Social Problems*, 54(4), pp. 434–453.

Skifter Andersen H. (2012) *The Importance of Housing Policies and Housing Markets for the Housing of Immigrants in the Nordic Countries*. Horsholm: Danish Building Research Institute at Aalborg University.

United Nations Department of Economic and Social Affairs 2015 report: *Trends in International Migrant Stock*. New York.

Whitehead C. and Scanlon K. J. (Eds.) (2007) *Social Housing in Europe*. London: London School of Economics and Political Science.

Index

Note: **Bold** page references indicate tables. *Italic* references indicate figures.